'Grow

ON CROP CULTURE

Geo. J. Ball™
PUBLISHING

GrowerTalks® on Crop Culture

Copyright © 1991 by Geo. J. Ball Publishing
1 North River Lane, Suite 206
P.O. Box 532
Geneva IL 60134-0532 USA

No part of this book may be reproduced, transmitted or stored in any form or by any means, electronic or mechanical, without prior written permission from the publisher.

ISBN 0-9626796-1-5

Printed in the United States of America.

10 9 8 7 6 5 4 3 2 1

Editor: D.J. Hamrick
Copy editor: Sherri M. Kangas
Production manager: Janet E. Sandburg
Cover design and illustration: John N. Landahl
Printed by Palmer Publications.

CONTENTS

Introduction xiii

Achillea

Achillea filipendulina
Culture notes, March 1990
   by *Leonard P. Perry* 1

Achimenes

Achimenes hybrids
Culture notes, April 1988
   by *Lin Saussy Wiles* 3

Alstroemeria

Alstroemeria
Culture notes, January 1987
   by *Mark P. Bridgen* 6

Alstroemeria culture, January 1984
   by *W. E. Healy and H. F. Wilkins* 8

Amaryllis

Amaryllis sp.
Culture notes, November 1989
   by *Xuri Zhang* 11

Anemone

Anemone coronaria
Culture notes, February 1988
   by *Meredith Shank* 13

Anigozanthos, Kangaroo paw

Kangaroo paw—a new crop for U.S. production, December 1986
   by *Mark S. Roh, Roger H. Lawson and Robert G. Anderson* 16

Antirrhinum, Snapdragon

Antirrhinum majus
Culture notes, September 1989
   by *Meredith Shank* 20

Aphelandra, Zebra plant

Aphelandra squarrosa
Culture notes, June 1985
   by *Teresa Aimone* 22

Astilbe
Astilbe x *arendsii*
Culture notes, October 1990
  by Jim Nau 25

Azalea, see Rhododendron

Bedding Plants
Sowings, December 1987 to November 1988
  by Nancy Bogart 27

Begonia, Fibrous
Begonia semperflorens cultorum
Culture notes, December 1990
  by Jim Nau 35

Begonia, Tuberous
Begonia socotrana x *Begonia tuberhybrida*
Culture notes, May 1987
  by Ron Adams 37

Brachycome
Brachycome iberidifolia
Culture notes, May 1986
  by Teresa Aimone 41

Caladium
Caladium x *hortulanum*
Culture notes, April 1985
  by Teresa Aimone 44

Calceolaria
Calceolaria hybrida
Culture notes, August 1986
  by Teresa Aimone 47

Campanula
Campanula isophylla
Culture notes, December 1987
  by Ron Adams 50

Chrysanthemums
Chrysanthemum frutescens
Culture notes, September 1990
  by Ron Adams 52

Chrysanthemums, garden

Chrysanthemum morifolium
Culture notes, May 1989
   by Ed Higgins 54

Growing ideas, September 1985
   by P. Allen Hammer 56

Codiaeum, Croton

Codiaeum variegatum, pictum
Culture notes, January 1985
   by Janet Langefeld 58

Cotula

Cotula turbinata
Culture notes, January 1989
   by Teresa Aimone 62

Crossandra

Crossandra infundibuliformis
Culture notes, February 1985
   by Janet Langefeld 63

Cut flowers

Perennial pickings: 5 garden favorites
cut from the greenhouse, March 1990
   by Richard R. Iversen 66

Cyclamen

Cyclamen persicum
Culture notes, October 1988
   by Teresa Aimone 69

Cyclamen persicum
Culture notes, June 1986
   by Teresa Aimone 71

Dahlia

Dahlia pinnate
Culture notes, December 1985
   by Teresa Aimone 75

Dahlia variabilis
Culture notes, March 1991
   by Sherri Neal 77

Delphinium

Delphinium sp.
Culture notes, August 1990
   by E. Jay Holcomb and David J. Beattie 79

Euphorbia, White top

Euphorbia marginata
Culture notes, January 1989
by *Teresa Aimone* 81

Euphorbia, Poinsettia

How to schedule and grow poinsettias for profit, May 1990
by *Harry K. Tayama, C.C. Powell and Richard K. Lindquist* 82

Euphorbia pulcherrima
Culture notes, April 1989
by *Carolyn Mack* 86

Bonzi and Sumagic on poinsettias
Growing ideas, August 1988
by *P. Allen Hammer* 88

Poinsettia variety evaluations
Research update, July 1988
by *C. Anne Whealy* 91

Eustoma, Lisianthus

Eustoma grandiflorum
Culture notes, July 1990
by *Daniel J. Jacques* 93

Exacum

Exacum sp.
Culture notes, April 1990
by *Jack Sweet and Paul Cummiskey* 95

Freesia

Fill out your sales with potted freesia, May 1990
by *Miriam Levy* 98

Cut freesias year-round, January 1986
by *Debbie Hamrick* 101

Fuchsia

Fuchsia x *hybrida*
Culture notes, November 1985
by *Teresa Aimone* 104

Gerbera

Gerbera jamesonii
Culture notes, June 1987
by *Lewis Howe* 107

Godetia
Clarkia amoena ssp. *whitneyi*
Culture notes, May 1990
    by Robert G. Anderson 109

Gomphrena
Gomphrena haageana and *G. globosa*
Culture notes, January 1990
    by Meg Williamson 111

Heliconia
Heliconia psittacorum
Culture notes, January 1986
    by Teresa Aimone 114

Hemerocallis, Daylily
Hemerocallis sp.
Culture notes, July 1989
    by Russell Miller 116

Hibiscus
Hibiscus rosa-sinensis
Culture notes, September 1985
    by Teresa Aimone 118

Hosta
Hosta sp. Frances Williams
Culture notes, June 1989
    by Russell Miller 121

Hydrangea
Hydrangea macrophylla
Culture notes, December 1984
    by Teresa Aimone 124

Hypoestes
Hypoestes phyllostachya
Culture notes, February 1987
    by Debbie Hamrick 129

Impatiens, New Guinea
Impatiens x *hybrida*
Culture notes, October 1989
    by Edward P. Mikkelsen 132

Impatiens wallerana

Impatiens wallerana
Culture notes, November 1990
  by Teresa Aimone 134

Kalanchoe

Kalanchoe blossfeldiana
Culture notes, October 1986
  by Teresa Aimone 137

Lilies, Easter

Bulb mites
Working the bugs out, October 1989
  by Mark E. Ascerno 139

Graphical tracking timetable, October 1988
  by Debbie Hamrick 141

Lilies, other

Lilium speciosum
Culture notes, December 1986
  by Teresa Aimone 144

Limonium, Statice

Limonium latifolium x *Limonium bellidifolium* Misty Series
Culture notes, April 1991
  by Fred Meyer 147

Limonium sp.
Culture notes, June 1988
  by Jim Nau 149

Mums, see chrysanthemums

Narcissus, Daffodils

Narcissus sp.
Culture notes, September 1986
  by Teresa Aimone 153

Nephrolepis, Boston Fern

Nephrolepis exaltata 'Bostoniensis'
Culture notes, February 1990
  by John Erwin 155

Nerine

Nerine bowdenii
Culture notes, October 1985
  by Teresa Aimone 157

Ocimum, Basil
Ocimum basilicum
Culture notes, March 1988
   by *Thomas De Baggio* 160

Otacanthus
Culture notes, April 1986
   by *Teresa Aimone* 163

Pardancanda
X *Pardancanda Norrisii*
Culture notes, December 1989
   by *Lin Saussy Wiles* 166

Pelargonium, Regal geraniums
Pelargonium x domesticum
Culture notes, October 1987
   by *Ron Adams* 168

Pelargonium, Zonal geraniums
Fast cropping
Culture notes, March 1989
   by *Joe O'Donovan* 171
Growing with Seeley, March 1984
   by *John G. Seeley* 173

Perennials
Perennial culture 101:
A handbook of the top six, March 1990
   by *Miriam Levy* 175
Perennial seed germination
Culture notes, November 1988
   by *Jim Nau* 179
Perennial production
A grower's notebook for summer, June 1988
   by *Miriam Levy* 182

Poinsettia, see Euphorbia

Primula, Primroses
Primula obconica
Culture notes, September 1988
   by *Teresa Aimone* 185
Primula acaulis
Culture notes, July 1987
   by *Ron Adams* 187

Primula, Primroses (cont.)

Primula acaulis: Culture, scheduling, cooling; October 1985
  by Vic Ball 189

Radermachera, China Doll

Radermachera sinica
Culture notes, December 1988
  by Ron Adams 192

Ranunculus

Ranunculus asiaticus
Culture notes, November 1987
  by Lewis Howe 194

Rhipsalidopsis, Easter cactus

A grower's guide to commercial production
of Easter cactus, November 1989
  by Thomas Boyle and Dennis Stimart 197

Rhododendron, Azalea

Rhododendron obtusum, Rhododendron simsii
Culture notes, August 1985
  by Debbie Hamrick 200

Rosa, Mini roses

Rosa sp.
Culture notes, February 1986
  by Teresa Aimone 204

Rosemary

Rosmarinus officinalis
Culture notes, August 1988
  by Thomas De Baggio and Thomas Boyle 206

Saintpaulia, African violet

Saintpaulia ionantha
Culture notes, February 1989
  by Russell Miller 210

Schlumbergera, Christmas cactus

Schlumbergera bridgesii, Schlumbergera truncata
Culture notes, July 1985
  by Teresa Aimone 212

Senecio, Cineraria

Senecio x *hybridus*
Culture notes, July 1986
   by Teresa Aimone 215

Sinningia, Gloxinia

Sinningia x *hybrida*
Culture notes, August 1987
   by Teresa Aimone 217

Smithiantha

Smithiantha zebrina
Culture notes, April 1987
   by Teresa Aimone 219

Solanum, Ornamental pepper

Capsicum sp., *Solanum pseudo-capsicum*
Culture notes, May 1985
   by Teresa Aimone 222

Streptocarpus

Streptocarpus x *hybridus*
Culture notes, September 1987
   by Lin Saussy Wiles 225

Trachelium

Trachelium caeruleum
Culture notes, July 1988
   by Lewis Howe 228

Torenia

Torenia fournieri
Culture notes, May 1988
   by Meredith Shank 230

Viola

Viola x *Wittrocklana*, Garden pansy
Viola cornuta, Horned violet
Viola tricolor, Johnny jump-up
Culture notes, June 1990
   by Jim Nau 233

Zantedeschia, Calla lily

New Zealand Callas, February 1988
   by A. C. Jamieson 235

Introduction

GrowerTalks on Crop Culture is a primer for growing everything from Achillea to Zantedeschia. Drawn from *GrowerTalks* magazine's monthly "Culture Notes" column, as well as other cultural articles, this handbook is filled with tips written by floriculture industry experts.

Learn the optimum soil, light, temperature and fertilizer for producing each of these major crops. Find out the best propagative methods and how to grow each crop to marketable size. Be forewarned and forearmed in preventing typical pests and diseases. Discover all the new varieties now available.

GrowerTalks on Crop Culture contains all the advice you need for producing many major bedding plant, foliage, bulb, potted plant and cut flower crops, in addition to a collection of perennial culture tips. Five separate articles highlight aspects of poinsettia production, including using growth regulators and evaluating varieties.

With ***GrowerTalks on Crop Culture*** in your library, you'll be well-informed and ready to start successful production programs. Now all you need is the greenhouse!

The Editorial Staff of *GrowerTalks* magazine.

Achillea

Culture notes

March 1990

by Leonard P. Perry

Yarrow
Family: Asteraceae
Genus, species : *Achillea filipendulina*

Rugged, quick and easy to grow, prolific and a long lasting bloomer (4 weeks is common), good cut or dried showy and non-invasive in the landscape, relatively pest free except for occasional aphids—all characterize Achillea, an herbaceous perennial. Fernleaf yarrow is so named from its deeply cut, gray-green and strongly scented leaves. Hardy in U.S. Department of Agriculture Zones 3 to 9 and native to Caucasus, its 3- to 5-inch mustard yellow flowers top 3- to 5-foot stems.

Achillea filipendulina is one of several species of ornamental yarrows out of about 100 total species. It's the related and widely grown species *Achillea millefolium*, common yarrow, from which "Achillea" comes. This species was grown in Europe as early as the 15th century as an herb to cure toothaches, as an additive to ale instead of hops, and as a medicine to heal soldiers' wounds. "Common" because of its tendency to spread rapidly and invasively, it's popular due to its many cultivars, from red and pink to white, and new pastel "earth" colors between.

Propagation

Propagate by seed, stem cutting or division. Division is perhaps the most used, least likely to lead to variability and is recommended for named cultivars. Division of 3-year or older plants can be done in spring or fall, cutting stems back ½ to ⅔ after dividing.

Sow seed at 65 to 70 F and expose to light for germination in 10 to 15 days. Transplant 3 to 4 weeks after sowing. If planting outdoors, 6 weeks may be required for 273 or similar plugs, 8 weeks for 6/8 cell packs. Transplanting from plugs to cells gives similar growth to direct cell sowing.

With good germination, plugs planted directly to the field or final pot are the least expensive. With poor germination, or if plugs cannot be planted directly, transplanting from broadcast flats or plugs to cells is the most cost effective method and the most efficient use of greenhouse space.

Achilleas flower the same year sown, but fernleaf yarrow flowers later. A vigorous grower, achillea grows best and most economically in the field, although container production is possible.

Achillea

Growing on

Field: Seedlings, divisions or bare root plants can be planted from spring on into most welldrained soils. Raised beds aren't necessary. Yarrows aren't particular as to pH (5.5 to 7), and low to moderate fertility is best. Too much fertility leads to excessive growth and the need to stake.

Remay, as a season extender, has little effect on growth, but black plastic ground cover may double the number of stems in addition to providing a weed barrier. Dacthal, Treflan and Fusilade 2,000 have been used successfully for chemical weed control—check labels before using. No overwintering cover is needed.

Container: Our research shows when a granular fertilizer is used (1 tablespoon 10-10-10 per 1-gallon pot), there are no growth differences between soil-based and soilless bark-based media, with best growth in a soilless peat-based medium. With a slow release fertilizer (Osmocote at lowest labelled rate), growth is similar in all media.

Seedlings, divisions or bare root plants can be potted anytime, although early spring gives saleable plants the first year. Many growers pot in mid to late summer, even early fall (propagating after the spring rush), with plants overwintered for next season sales.

Some cover is usually needed over pots except in warmer areas. An overwintering house (Quonsets are most common with one or two layers of white poly in the far North) at 30 to 35 F is ideal, especially if mice live nearby, although foam materials and other methods are common. Most effective at insulating and moderating temperatures and least expensive are two layers of poly (white on top) with 1 foot fluffed straw between. Apply covers before first heavy snow and remove as soon as possible in spring.

Cut flowers

Grown as a field cut flower, reliable yields can be expected from the second year on—30+ stems per square foot (plants on 1-foot centers) in the North to 40+ stems per square foot in the South for Coronation Gold achillea. Further spacing results in more stems per plant but fewer per area (and more stems per area result in greater returns).

Fernleaf yarrow flowers last longest and retain their color when dried. Common yarrow cultivars may fade when older, especially darker colors.

Cut before pollen appears, with stems as long as possible, in early morning or late evening. Place immediately in water. If held over several hours, leave stems in neutral pH tap water (over distilled water) to add two days. Moving stems to a preservative solution may add 2 to 4 days to vase life (expect 9 days or more total).

Cultivars

Gold Plate, Cloth of Gold and Parker's Variety are all similar cultivars of *Achillea filipendulina*, growing 3 to 5 feet. Often listed under this species, but actually a cross of *A. filipendulina* and *A. clypeolata*, is Coronation Gold—best propagated by division, as seed is often from *A. filipendulina*.

As mentioned, *Achillea millefolium* includes many different cultivars—Cerise Queen, Red Beauty, Fire King and Fire Beauty are similar with red flowers; Rose Beauty has rose flowers; White Beauty has white flowers; the 1990 All America Selections winner, Summer Pastels, is a mix of salmon, cream, light pink and yellow flowers.

Popular hybrids include: Galaxy series with red or pastel flowers—crosses of *A. millefolium* and *A. taygetea* and Moonshine with sulfur yellow flowers and gray-green foliage—a cross of *A. clypeolata* and *A. taygetea*.

"

Leonard P. Perry is extension assistant professor, University of Vermont, Burlington.

Achimenes

Culture Notes

April 1988

by Lin Saussy Wiles

Achimenes
Family: Gesneriaceae
Genus, species: *Achimenes hybrids*

Achimenes are heirloom pot plants that were much appreciated during the 1800s for their beauty and are now enjoying a resurgence of interest with the many fine new cultivars since 1940.

Available cultivars offer a full spectrum of flower colors and the leaves also add a note of drama with colors ranging from light green to dark bronze. They offer a grower the opportunity to cut shrinkage losses to a minimum because

Achimenes

plants can be sold in spring, throughout the summer, and some can be held over winter for sales the following year.

Propagation

Seed: *Achimenes* seed is very fine, 3 million seed per ounce, and should be sown thinly in market-packs. Seed can be sown anytime but the best results are obtained with December to March sowings. Sow in a porous, well-drained medium, such as a peat-lite mix; do not cover seed; moisten thoroughly and keep moist. Germination takes 14 to 21 days at 75 to 80 F soil temperature. Plants grown from seed bloom within five to seven months.

Clones: Rhizomes, or rhizome scales (broken-up rhizomes), produce flowering plants in two to four months. They can be started as early as December or as late as June, but are normally started February through April. Rhizomes should be planted horizontally, 1 inch apart and $1/2$ inch deep in either flats, cell-packs or their final pots. Most cultivars require four rhizomes for a 4-inch pot, or six rhizomes for a 6-inch pot, while base-branching cultivars require fewer rhizomes per pot. Water immediately after planting, then water just enough to keep medium barely moist until sprouts appear. Overwatering at this stage can cause rot. Sprouting occurs in three to six weeks of 60 to 75 F nights. Warmer nights hasten sprouting. After sprouting, temperature can be held at 60 F and the plants must be kept evenly moist.

Cuttings can be taken from the time plants are pinched back until they begin to go dormant. Rooting occurs in 10 days. These are less vigorous than seed or rhizome propagated plants but bloom in three months or less depending on the maturity of the plant at the time the cutting is taken.

Growing on

Transplanting: In six to eight weeks from sowing, seedlings will have two true pairs of leaves. Transplant into 72-pack cells or a similar container using a well-drained peat-lite medium. When plants are established, in four to six weeks, they can be repotted directly into containers for sale.

Plants grown from rhizomes should be transplanted when they are 2 inches tall, while those from cuttings can be transplanted when they are well-rooted.

Lighting: Summer shade is required since direct sunlight causes leaf scorching and flower fading. The greater the amount of bright filtered sunlight available, the more floriferous the plants will be. Good results can be obtained with up to 5,000 footcandles.

Temperature: Tolerance tends to be towards very hot growing conditions. Plants flower somewhat less in extreme (100 F) heat but recover well when temperatures moderate to the ideal range of 75 to 90 F days and 65 to 70 F nights. They can tolerate temperature ranges of 50 to 120 F. Proper ventilation is essential for healthy, well-developed plants.

Fertilization: After transplanting, fertilize with 20-20-20 at 200 ppm N on a constant feed schedule to produce lush, floriferous plants. If some plants are held throughout the summer for "instant color" sales, alternate with 15-30-15 to further boost flowering. Fertilization should be continued until plants start yellowing as dormancy approaches.

Watering: Irrigating with warm water (68 to 76 F), instead of normal ground temperature water, will produce the best results, and ideally irrigation should be done in the mornings. Once plants are actively growing they must be kept evenly moist, excessive dryness can induce early dormancy.

Special care: Spacing should be monitored as plants will stretch rapidly if held too close for very long. Four-inch pots set on 10-inch centers do well.

Pinching can be used to moderate height, increase plant fullness and to obtain additional plant material. Non-branching cultivars can be made to branch by pinching terminal growth when plants are 3 to 4 inches tall and continuing to pinch plants for several weeks thereafter.

Growth regulators may be advisable for some cultivars. B-Nine (5,000 ppm to 7,500 ppm) as a single spray when shoots are 2 to 3 inches long can be effective.

Dormancy begins in late summer or early fall when plants start to yellow and flowering slows and finally ceases. At this point, watering frequency should be gradually reduced and fertilization should be discontinued. Cut back plants when stems turn brown and set aside in an area above 50 F. Dormancy lasts from two to five months depending on the cultivar.

Insects and diseases: Botrytis, a problem where there is insufficient ventilation or spacing, can be relieved by using Benlate. Whiteflies are occasionally a problem—Orthene, Resmethrin or Safer's Insecticidal Soap sprays alternated every 10 to 14 days can be used for localized infestations. Pollution can be a problem for a number of cultivars, especially where there is a high level of air pollution. Testing and selection of cultivars for large city sales is recommended.

"

Lin Saussy Wiles was a plant breeder for Park Seed, Greenwood, South Carolina when this article appeared.

Alstroemeria

Culture Notes

January 1987

by Dr. Mark P. Bridgen

Alstroemeria
Family: Alstroemeriaceae
Genus: *Alstroemeria*

Alstroemeria are traditionally grown for cut flowers; however, in 1985 a dwarf variety called Rosy Wings was introduced to the United States. Rosy Wings makes an unusual and attractive 6-inch pot crop, and shows tremendous potential for becoming popular with both the grower and the consumer. The individual flowers have three pink outer petals and three white inner petals with brown specks. Many people confuse them with orchid flowers because of their showy size and color. The flowers can be removed and used as cut flowers. Rosy Wings is not protected by grower agreements or patents and is therefore free to propagate.

Propagation: *Alstroemeria* Rosy Wings can be propagated by seed, but variability exists in the seedlings and it takes four to eight weeks for the seed to germinate. *Alstroemeria* are best propagated by rhizome division. The plants can be divided any time after flowering stops from September to mid-November.

Three rhizomes should be planted per 6-inch pot with one to three growing shoots per rhizome. Pot the rhizomes toward the center of the pot with growing end outward. While the rhizomes should be planted very shallow, they should not be exposed on the surface. The large storage roots that accompany the rhizomes may be planted at any depth. Once planted, any dead or damaged plant material should be removed. The pots should be drenched with a fungicide to prevent root rot and placed in a cool house with 55 to 65 F night temperatures. The plants should be fertilized with 100 ppm nitrogen for the first few weeks.

Rhizomes of Rosy Wings can be purchased through Holland Park Products in Marietta, Georgia (800) 334-8007.

Cold Treatment: Rhizomes should be allowed to grow vegetatively for four to ten weeks after potting. Then, they need to be cooled in a refrigerator at 40 F or in coldframes that do not get below freezing. The foliage should not be removed before the cooling period. Pots should be drenched with a fungicide before cooling and then allowed to dry moderately. Once in the refrigerator or coldframe, the plants should be kept on the dry side, but never allowed to completely dry.

Growing-on: Pots of Rosy Wings should be removed from cold treatment at least 11 weeks before they are to be sold. Any shoots that are damaged or that

have Botrytis should be removed. Pots should be placed in cool houses with 55 to 65 F nights and 58 to 70 F days. One to two weeks after removing the pots from their cold period, one-fourth of the shoots should be cut back to the ground. This allows more light to the leaves and will give better flowering later. If the shoots are tall for the pot, pinch off the top halves of all shoots. This will encourage new shoots and will help dwarf the plant. About one month before sale, stems that have died should be removed. If the pots are thick with shoots, one-fourth of the stems should be removed to ground level. This will also help shape up the plant.

General culture

Greenhouse medium—pH: Any soil-based or soil-less greenhouse medium can be used for *Alstroemeria* if it is well-drained. The pH is best at 5.4 to 5.8.

Fertilization: *Alstroemeria* can tolerate high fertilization levels, but they only need 200 to 250 ppm nitrogen on a weekly basis. They should only be fertilized with 100 ppm nitrogen for the few weeks immediately following division.

Temperature: One of the advantages of *Alstroemeria* is that they prefer cool temperatures. Nights of 55 to 60 F and day of 58 to 70 F are ideal. Warm temperatures may cause the plants to stretch and flower buds to abort.

Light: Light intensity and duration both play an important role in the flowering of *Alstroemeria*. High light intensity is essential for flower induction and development. Unless supplemental lighting is used, Rosy wings will not flower adequately in low light regions before mid-March. Long-days should be given to *Alstroemeria* once they are removed from the cold period; this will hasten flowering. If light intensities remain low, flowering will be inhibited and plants will stretch.

Pinching and thinning: If pinching and thinning are used correctly under proper light, potted *Alstroemeria* can remain dwarf without growth regulators. If pots are thick with foliage, one-fourth of the spindly and older shoots should be removed to allow more light to the plant. If plants are tall, they can be pinched 1 to 2 inches below the desired height and new flowers will remain dwarf. Pinching and thinning should be done carefully four to six weeks before sales.

Diseases and insects: There are no major insect problems with *Alstroemeria*, and root rots are the only major disease problems. Regular fungicide drenches at the time of division and refrigeration, and good greenhouse technique should prevent any major problems.

Post-harvest care: One of the biggest selling points to customers may be that potted *Alstroemeria* will continue to flower until hot temperatures arrive. They can remove the flowers and enjoy them as cuts, or they can plant the whole pot outside and enjoy them in the garden. Rosy Wings should not be planted outside until the danger of frost is past. If desired, the customer may want to dig up the rhizomes in the fall, store them over winter in a cool, but not freezing, location and then plant them outside in spring.

"

Dr. Mark P. Bridgen is assistant professor of floriculture, University of Connecticut, Storrs.

Alstroemeria culture

February 1984

by W. E. Healy and H. F. Wilkins

The new alstroemeria cultivars that are available from Europe are the result of years of interspecific breeding and irradiation to induce mutations of various cultivars. The majority of the new hybrid cultivars have originated from the van Staavern Company (The Netherlands) and Parigo Seed Company (England). Recently the Wulfinghoff Company (The Netherlands) has introduced several new and interesting alstroemeria cultivars. Plants are normally leased from the breeder; a yearly royalty is assessed on the square footage in production. The Fred Gloeckner Company is the U.S. representative for the van Staavern Company.

Alstroemeria species and hybrids come in many different shades of reds and yellows. The native species are found from the snowline of the high mountain plateaus of the Andes in South America down through the highland forests to the coastal deserts. The requirements for floral induction in the hybrids can be separated into two groups; the white/yellows, and the red/oranges, the latter of which originated from several different interspecific crosses. The white/yellows require a shorter cold period, a higher "devernalization" temperature, and a shorter photoperiod for flowering than the red/orange group.

Plant characteristics

As the leaves on an alstroemeria shoot unfold, they rotate 180 degrees so that the adaxial (upper) surface becomes the abaxial (under) surface. The inflorescence consists of a whorl of flowering cymes with each individual cyme bearing one to five sympodially arranged (arranged as in cattleyas and cymbidiums) flower buds. A cyme is a type of infloresence which is usually broad, and more or less flat topped. The central or terminal flower opens first on a cyme.

The alstroemeria plant consists of a white fleshy rhizome from which arise aerial shoots and a moderately fibrous root system which can become thickened like a dahlia. The growing points of the rhizomes give rise to aerial shoots. Each new aerial shoot arises from the first node of the rhizome, and a lateral rhizome develops in the second node of the aerial shoot.

The aerial shoots can be either vegetative or reproductive. Normally, shoots that have unfolded more than 30 leaves will not flower and are vegetative. Once rhizomes are induced to flower by low temperatures, all shoots that form will flower until plants become "devernalized" by high temperatures.

Flower induction

The flowering control mechanism for alstroemeria hybrids appears to be of a 2-phase nature with a primary cold temperature requirement and a secondary photoperiod requirement. As stated earlier, the temperature effect shows a group specificity with the white/yellow group requiring a shorter cold treatment (two to four weeks of 40 F) for floral induction than the red/orange group which needs four to six weeks of 40 F for floral induction.

Once flowering begins, the plants will continue to produce flowering shoots indefinitely until the soil temperature rises above 60 F for extended periods of time. When plants were grown at continuous 55 F soil temperature, the plants continued to flower indefinitely regardless of air temperature and regardless of photoperiod. Since the below ground part of the plant must be kept cool (55 F) for continued flowering, deep soil mulches or misting the mulch to encourage evaporative air cooling will help maintain a cool soil temperature. This can be used in conjunction with evaporative air cooling during periods of warm temperature.

The other component that controls flower induction is photoperiod. Once plants have perceived an adequate cold treatment, a 13-hour photoperiod, as obtained by using standard chrysanthemum lighting as a night interruption, hastened floral initiation. Lighting non-vernalized plants will not induce flowering. Photoperiods longer than 13 hours will not promote any earlier flowering, but may decrease flower production. We have found that a 4-hour night interruption (incandescent source, 10 p.m. to 2 a.m.) is adequate. Another method we use: Add the length of normal existing daylength plus the number of hours of night interruption needed to equal 13 hours. These night interruptions are effective in promoting earlier flowering without fewer flowers.

Lighting should occur from about September 1 to April 15 at 45 N latitude (St. Paul, Minnesota). Check with your local weather bureau to determine the exact dates when the daylength is less than 12 hours at your latitude.

Light intensity has been shown to affect flower development. In northern Europe where light intensity in the winter is significantly reduced, bud blasting is a problem. In Minnesota we have only occasionally observed bud blasting, and this may have been related to soil temperatures which were too cool. Since the number of cymes per stem is positively correlated with the stem diameter, alstroemerias should be grown with the maximum available light so that the greatest number of cymes per stem can be attained.

Propagation

Alstroemeria plants should be divided every second or third year depending on the cultivar and growth characteristics. About one to two weeks prior to dividing, plants should be severely pruned leaving only the youngest 6- to 8-inch shoots. The pruning will facilitate handling. When digging well-established plants after flowering decreases in late summer or early fall, care should be taken to dig deep enough to get the growing point as the rhizome can grow 12 to 14 inches deep. Each new division should consist of a single rhizome with an undamaged, blunt growing point, some new aerial shoots, and, most important, some large fleshy storage roots.

The presence of these thickened storage roots is critical for the re-establishment of the plants since the first new roots will arise form these enlarged storage roots. Normally, only the youngest 1 to 3 inches of rhizome is kept, with the older portion of the rhizome being discarded. These older rhizomes are of no value as the lateral rhizomes that may arise from them are weak and do not appear to regain vigor.

Immediately after the rhizomes are divided or new rhizomes received, they should be planted. It is essential that pots, soil, or ground beds be ready before plants are divided or received. Normally, extra plants are potted up to replace plants that die or are not as vigorous as others. It is expected that 5 to 25 percent of the plants will not survive transplanting. To increase the survival rate, a fungicidal drench (8 ounces each of Lesan and Benlate per 100 gallons of water) is recommended at the time of planting, and again a month later if vigorous root growth is not observed. Excess watering will quickly rot the rhizomes. After the initial watering with the fungicidal drench, spot water plants as they become dry. Grow the plants at 65 F until they become well established (four to eight weeks), before lowering the temperature to 40 F.

Harvesting and marketing

When new growth begins, many shoots will form. Removing some of the weak vegetative shoots has shown to increase flower production. Pinching plants encourages growth or lateral branching of the lateral rhizomes. Shoots should be pulled since cutting the stem and not pulling them off the rhizome will decrease flower production. A quick, upward pull will cleanly remove the shoots from the rhizome. Stems are harvested when the first group of flowers begins to open. Care should be exercised with young or poorly rooted plants since the rhizome may be uprooted or torn loose from the soil if careless stem removal occurs.

There are currently two methods used for marketing alstroemerias. One method is simply to place 10 to 12 flowers in a bunch. The more desirable method grades stems into three grades. Grade 1 has stems with five or more cymes per 36-inch-plus stem length, Grade 2—three to four cymes per 24-inch-plus stem length, and Grade 3 has less than three cymes per 12-inch-plus stem length. Stems are downgraded for crooked or damaged stems or poor foliage. A strong 3 to 4 layer support system is recommended. Alstroemerias can produce shoots up to 6 feet tall.

Extra tips

Alstromerias are heavy feeders. We have fed up to 600 ppm nitrogen in the form of potassium nitrate twice a week to vigorous plants without any adverse affects. Ammoniacal forms of nitrogen fertilizer should be avoided, as at 55 F growing temperatures, ammonia is not readily converted to nitrate. Application of minor nutrients may be required to maintain optimum levels within the plant.

Insects and diseases are essentially not a problem on alstroemerias. Whiteflies and aphids may appear during warm weather. We have observed leaf mottling and verified this as a virus infection.

”

W.E. *Healy and H. F. Wilkens were with the University of Minnesota, St. Paul when this article appeared.*

Amaryllis

Culture notes

November 1989

by Xuri Zhang

Amaryllis
Family: Amaryllidaceae
Genus, species: *Amaryllis* sp.

Amaryllis, or Barbados lily, is a very old and attractive plant—the first amaryllis species was recorded in scientific literature 300 years ago in 1689; yet its commercial production is still limited to a small scale.

Most amaryllis species are natives of tropical and subtropical Americas, with the largest number of species native to the Amazon River Basin in Brazil, Bolivia and Peru. Flowers occur in a variety of sizes, shapes, colors and fragrances. Some blooms are double or huge and open-faced; others mimic irregular orchid shapes. Plants also vary greatly in height, from 20 centimeters to more than 70 centimeters.

Propagation
Seed. Collect seed from capsules and air-dry 24 to 48 hours at room temperature if they are to be stored. Best germination occurs when seed is planted immediately after harvesting. Seed germinates in seven to 14 days, but may take longer for the first leaflet to show above the soil surface. After sprouting, feed seedlings regularly every two weeks with a complete, mixed fertilizer.

Transplant seedlings to 3-inch pots when they have two or three leaves. Thereafter, transplant seedlings to the next larger size pot as they become rootbound, until reaching a 6- or 8-inch pot. From germination to mature bulb takes 18 to 24 months, depending on cultivar and environmental conditions.

Vegetative propagation is done two ways: natural increase by offsets from mother bulbs or increase by cutting methods that stimulate bulblet formation. Some species and hybrids make offsets profusely, while others rarely, if ever, produce any. For commercial production don't rely on offsets except where they are produced in abundance. Separate offsets from the mother bulb after they have a separate root system.

Usually the large flowering Dutch hybrids and similar American hybrids don't make offsets in sufficient quantity for commercial use. Propagate them by cutting using bulb-scale, fractional scale-stem and bulb-base removal.

The bulb-scale method, carried out from May to September, separates and sprouts bulb scales that have the power of regenerating bulblets. Place scales in a slightly slanting position in a flat. In three months, pot bulblets. Keep plants in a greenhouse at a soil temperature of 77 to 82 F after potting until they become sufficiently strong.

For the **fractional scale-stem method,** wash the bulb, trim leaves to the neck, cut roots back to an inch in length, then cut the bulb lengthwise into two, four, eight or 16 fractions. Cut each fraction into two to six fractions, as long as each has two or more bulb-scale parts attached. As a rule go for 16 final fractions for sprouting.

Place cuttings in a slanting position in a flat. Regulate water carefully; excess moisture will cause rot. In six weeks sprouts form, and leaves show in three months. Summer cuttings produce bulblets ready for transplanting the following February or March.

In **bulb-base removal method,** slice off the bulb base just below the scales and dust the exposed surface with IBA-rooting promoter. Store bulb in the shade. After eight to nine weeks, when a ring of tiny bulblets has appeared around the base, set the mother bulb in a gravel hydroponic bed and feed regularly with a standard nutrient solution. Many bulblets reach a diameter of $1^{1}/_{4}$ to $1^{1}/_{2}$ inches in eight months.

Culture

Flower scapes initiate and develop in the bulb months before appearing. Care for plants to form the maximum number of immature inflorescences.

Choose bulbs undamaged by low temperature exposure and free from disease and insects. Bulbs should be of blooming size and well-grown if immediate flowering is expected. Consider bulb sizes at all times in relation to species or variety. Some hybrid amaryllis, particularly some miniature ones, have medium to small-size bulbs and still give excellent performance in sizes even below $2^{1}/_{2}$ inches in diameter.

Bulbs should have original roots intact. Bulbs with all roots removed to the root base are inferior during the first year and possibly also the next. They bloom if they are of flowering size, but quality will be impaired because nutrients cannot be taken up readily, and stored food reserves will be depleted.

Soil. For pot production use a medium made up of coarse sand, peat moss and soil. Keep soil slightly acid to neutral (pH 6.1 to 7.0), although plants will grow on very slightly to mildly alkaline (pH 7.1 to 7.8) soils.

Watering and fertilizing. Water amaryllis thoroughly when needed, but don't allow water to stand for long periods, except with severely root-bound plants. Apply a complete, mixed fertilizer every two or four weeks.

Pest control. Amaryllis is surprisingly free from fungal diseases, particularly when given optimum culture. As a general sanitary measure, remove all declining and dead leaves and flower stalks. The most persistent, destructive pests of amaryllis are mealybugs, red spider mites, scales and thrips. To control mealybugs and scales, spray insecticidal soap when needed. Dycarb spray can control thrips, and Pentac flowable will get rid of red spider mites.

Temperature and light. Since Amaryllis is a tropical or subtropical genus, plants respond to high temperatures both day and night. We suggest a minimum 70 F day and not lower than 60 F at night during the growth cycle. Temperatures lower than 50 F can be detrimental.

The area in South America where most species are indigenous is only a few degrees latitude south of the equator and usually at high altitude (4,000 feet or more)—areas subject to intense sunlight. Grow under full sun from sunrise to sunset. Bulbs grown under full or one-half shade may thrive but may bloom erratically. In full sun bulbs produce scapes each year.

"

Xuri Zhang is a research assistant in the Department of Horticulture, Pennsylvania State University.

Anemone

Culture notes

February 1988

by Meredith Shank

Anemone
Family: Ranunculaceae
Genus, species: *Anemone coronaria*

Anemone, a member of the buttercup family, is traditionally known as wind flower or lily of the field. The genus contains numerous species, some of which make excellent garden perennials. The vast majority of anemones grown today are developed from the species *coronaria* and sold as cut flowers. The new F_1 hybrid series, with large blooms on long strong stems and increased productivity, now dominate the European anemone market, where they bring premium prices. Peak production is from December through March. Hybrid anemones come in an array of colors, from deep blues to pink, red, white, wine, orchid and bicolors. Previously found only as mixes, separate shades are now available from some sources.

Propagation
Seed: Now available from a number of suppliers. There are 56,000 seed per ounce. Sow seed in mid-March to mid-April in a well-drained peat-based mix. Seed should be planted ½ inch apart. Anemones have roots that need room to

Anemone

grow. Cover seed lightly (1/8 inch) with soilless mix and use a Lesan/Benlate drench to water the seed in (1 1/2 teaspoons of each per 1 gallon of water).

Germinate at 59 F. Higher temperatures will reduce your germination. Maintain high humidity during germination either by misting, or by covering the flats with newspapers, sterilized burlap or with clear plastic that has been perforated with holes.

Plugs (1-inch or larger liners): The new F_1 hybrids are best grown from large-celled plugs. Many growers find that bringing in large (1- to 1 1/2-inch) plugs from specialists greatly simplifies growing and increases success. Plugs also reduce crop time significantly and allow you to bring plants in after the summer heat is over. Most prefinished plugs are delivered in late summer for first cuts in late November.

Growing on

From seed: Transplant eight to nine weeks after sowing when the seedlings are about 1 inch long, into cells that are 1 inch in diameter. Care should be taken to prevent injury to the delicate root system during transplanting. Use a well drained mix. Add 2- to 2 1/2 pounds of superphosphate per cubic yard of mix. It is critical that soil pH remain neutral, pH 6.8 to 7.0, throughout the growing season. Grow seedlings for about two months in the tray before transplanting into the final bed.

From transplant: In the ground bench, space plants out at 10 inches between rows and 6 inches between plants. The soil mix should be steam-sterilized and have a pH of 6.8 to 7.0. Finally, put a layer of sterilized straw down to keep the soil temperatures cool during the summer. Regardless of growing method, do not set the plants too deep into the final container. It is especially important not to cover the crown.

Shading: In areas of extremely high temperatures or high light intensity, heavy shading may be required. Shading should be removed in areas where cooler temperatures and cloudy weather persist throughout the growing season. In most areas, plants should be grown with maximum light available from September onward while still maintaining cool growing conditions.

Temperature: Anemones are at their best grown cool. Maintain 42 to 48 F night temperature and 58 to 65 F day temperature. During warm weather when cooler night temperatures are not possible, plants do better when grown in houses equipped with fan and pad cooling. Anemones require less labor than many other cool cut flower crops.

Fertilization: In order to stimulate root growth, immediately after transplanting, fertilize once with 9-45-15 (at a rate of 1 pound per 100 gallons of water and apply at 1 gallon per square foot) or any starter solution that will help to establish root development.

To grow-on, use a low ammonia or urea nitrogen fertilizer such as 15-16-17 or 20-19-18 at 200 ppm. Leach any salt buildup by irrigating with clear water every third watering.

Insects: Oxamyl or Malathion have been used successfully to control aphids (allow plants to become established before applying Oxamyl). Resmethrin aerosol can be used to control whiteflies. To control thrips, harvest and destroy the flowers and spray plants with Dursban.

Disease: To help control *Botrytis cinerea*, products such as Botran, (1/2 pound per 100 gallons of water), Ornalin, (1 pound per 100 gallons of water) or Daconil (1 1/2 pounds per 100 gallons of water) can be used safely on a monthly basis when plants are well established. Alternate the above mentioned chemicals and use a spreader/sticker agent. Good ventilation will help to control Botrytis.

Powdery mildew can be controlled using sulfur dust. Rhizoctonia can be controlled by a Terraclor drench. Subdue is effective in control of Pythium.

Again, good cultural practices—avoid overwatering and poor drainage—will reduce the chance for root diseases. During periods of high heat and humidity, spray every seven to 10 days with 1 pound Ferban plus ½ pound Benlate in 100 gallons of water to prevent leaf curl fungus.

Cultivars

Mona Lisa: Recently developed, Mona Lisa is the first and only F_1 hybrid anemone commercially available and the only series available in a full range of separate colors. Huge 4- to 5-inch blooms on 17- to 18-inch stems. Highly productive; dominates the European anemone market.

DeCaen: Developed in France in the 1800s. Single-flowered and usually seen as a mix, although some separate colors are available.

St. Brigid: Semi-double blooms on 12- to 15-inch stems. Separate shades and a mix.

,,

Meredith Shank is with PanAmerican Seed Co., West Chicago, Illinois.

The kangaroo paw—a new crop for U.S. production

December 1986

by Mark S. Roh, Roger H. Lawson, and Robert G. Anderson

The unusual and long-lasting flowers of the kangaroo paw make this Australian perennial a natural for cut flower and pot plant uses in the greenhouse industry. At present, there is some culture of this herbaceous plant in California, Australia, Israel, Netherlands and New Zealand. Since 1983, kangaroo paws have been evaluated as part of the New Crops Program at the United States Department of Agriculture, Agricultural Research Service, Florist and Nursery Crops Laboratory. The purpose of the research is to investigate factors that control flowering and to determine production practices for this new greenhouse crop.

Kangaroo paws are an unusual group of Australian wild flowers form the genera *Anigozanthos* (11 species) and *Macropidia* (one specie). The plants are herbaceous perennials with creeping, clumping rhizomes and basal, strap-shaped or sword-shaped leaves. The leaves are arranged in fans that resemble rhizomatous iris such as the common, tall, bearded types. Flowers with bright red, purple, green, or yellowish colors are borne in a single row along a tall, narrow, woolly spike or raceme that stands above the foliage. As the flowers open, the spike is bent and resembles a kangaroo's paw; thus, the common name. Kangaroo paws are monocots in the *Haemodoraceae*, a family closely related to the lily *(Liliaceae)* and iris *(Iridaceae)* families.

Natural habitat

Southwestern Australia is the natural home of the kangaroo paw. Species of *Anigozanthos* and *Macropidia* occur as part of the scrub vegetation along the coast and up to 100 miles inland. They grow in full sun among the shrubs and herbs common in open hearth and woodland areas where the soils are highly leached and mineral-deficient and in deep sands of the coastal plain.

Kangaroo paw species experience a warm temperate Mediterranean climate with summer drought and dominant winter rainfall (May to October inclusive) with a range from 15 to 48 inches according to species. In the North, summer temperatures average 80 F or more, with relatively frequent temperatures in excess of 100 F. To the south, temperatures are more moderate. Humidity is relatively low throughout the year.

All species are inactive from mid-summer, with some species showing a full dormancy, surviving solely as sub-surface rhizomes. Active vegetative growth resumes with lower temperatures and the onset of rains in autumn or early winter. Each species has a more or less distinctive flowering period with the range in late winter or early summer.

A. Humilus is more dwarf than others, with flower stalks reaching a height of up to 18 inches. Flowers are yellow to orange-red (tips).

A. manglesii is the official floral emblem of the State of Western Australia. Stems reach up to 3 feet tall. Stalks are covered with red woolly hairs that continue to the base of flowers. The remainder of the flower is green and hairy.

Description of the species

Twelve species of kangaroo paws are recognized and fall into two sections of the genus *Anigozanthos* together with the monotype *Macropidia fuliginosa*. It is important to review the native species so that the range of plant size and flower color can be evaluated for future use in cultivar development and production.

Angiozanthos manglesii: The red and green kangaroo paw produces 12- to 24-inch gray-green, broad, sword-shaped leaves with stiff hairs scattered on both leaf edges. The flowering stem emerges from the center of the leaf fans and grows up to 3 feet tall. The unbranched, or occasional branched, flower stalk is covered with scarlet-red woolly hairs that continue onto the swollen red base (ovary) of the flowers. The remainder of the 3-inch flower perianth is green and covered with dense greenish hairs. This species flowers from July to December, with peak flowering in September and October. *A. manglesii* is the official floral emblem of the State of Western Australia.

Anigozanthos bicolor: This red and green kangaroo paw grows 12 to 18 inches tall with narrow leaves and resembles a dwarf *A. manglesii*. It produces flowers with a red and green corolla from August to November and becomes dormant in summer.

Anigozanthos gabrelae: This is another red and green kangaroo paw closely related to *A. bicolor*, but of even dwarfer habit. Leaves rarely exceed 4 inches long with flower stems 8 inches tall. Flowering period is similar to *A. bicolor*.

Anigozanthos rufus: the red kangaroo paw grows on the southern coast from the Stirling Range to Cape Arid. Leaves are dull green in color; flowers with a deep red corolla are produced on 2- to 3-foot inflorescence from October to January.

Anigozanthos pulcherrimus: This species, known as golden kangaroo paw, is found in a restricted area on the west coast north of Perth. Leaves are gray-brown, and the compound inflorescence is 2 to 3 feet tall. Flower color ranges from rich yellow to full gold, rarely apricot, when flowers appear from November to February.

Anigozanthos viridis: The green kangaroo paw is found in wet to swampy areas on the west coast from August to Watheroo. The branched or unbranched flower stalk is almost 2 feet tall and is produced from September to November. The flowers are nearly 3 inches long and have a yellow-green base and bright emerald green corolla.

Anigozanthos flavidus: The Albany kangaroo paw has evergreen leaves; it grows in the cool and cloudy region from Albany to Cape Naturalisite, where rainfall is nearly 60 inches annually. The branched inflorescences are 5 to 6 feet tall and are produced from October to December. Flowers are over 1 inch long and range from green to predominantly deep red, due to the dense branched hairs on the exterior, which contrast strongly to the gray throat. A pink variant of this species is also found.

Anigozanthos humilis: The cat's paw is widespread, ranging from the south coast near Albany to the Murshism River. Leaves are short (up to 6 inches long) and green with stiff hairs. The branched or unbranched spike is 6 to 18 inches tall and bears yellow to orange-red flowers from June to November. This species has not grown extremely well in cultivation.

Macropidia fuliginosa: The black kangaroo paw grows near the west coast from Muchea to Geraldton. It grows in large clumps; the yellow-green leaves have a yellow marking with dark tip. The compound inflorescence grows up to 3 feet tall and is covered with black, branched hairs. Flower buds are also clothed in black hairs, but open to show greenish to whitish petals. The flowering season is from September to November.

Three additional species of *Anigozanthos* have been used in selection and breeding programs. These three species are all very localized in occurrence. *A. preissii* typically grows as single plants rather than clumps. The leaves are shiny green, narrow, fleshy, and grow approximately 18 inches tall. The flowers have a reddish-orange base and a yellow corolla. *A. onycis* and *A. kalbarriensis* are small plants with foliage 4 to 8 inches long and flower stems 6 to 12 inches tall. Flowers of *A. onycis* are red and yellow, and flowers of *A. kalbarriensis* are yellow and green.

Kangaroo paw selections

Kangaroo paws have been used in the landscape, for commercial cut flowers, and more recently as pot plants. Breeding and selection programs are continuing to develop plants for these uses.

Species such as *A. flavidus, A. manglesii, A. rufus,* and *A. pulcherrimus* are desirable green plants in Australia. However, only *A. flavidus* is commonly available from nurseries. This species has proved hardy, reliable, and long-lived in cultivation in southern temperate Australia with negligible susceptibility to pathogens. Clumps are multiplied and established easily. Rapid propagation, simple cultural needs, and low labor demand for maintenance have been the breeding program goals for landscape use of kangaroo paw.

Flowers of several species have been cut from the wild for many years, but cultivated cut flower production is increasing in both western and eastern

Australia. Presently, commercial plantings have been mainly seedling plants of *A. manglesii* and invitro-propagated clones of a few species. However, these are being replaced by new hybrid cultivars. Due to the variability of flower color, productivity, and inconsistent flowering of seed populations, breeding programs that produce suitable cultivars or relatively uniform seed lines are necessary for the future of commercial cut flower production.

Interest in the potential of kangaroo paw as a pot plant has developed. Mr. Mervyn L. Turner, Bush Gems Garden Nursery, Monbulk, Victoria, Australia, views dwarf and small-statured kangaroo paws as having great potential for containerized production and use. Dwarf kangaroo paws as container plants should resist warm and dry conditions and be suitable for patios in warmer climates, as well as for indoor use in northern areas. Bright flower colors and disease resistance found in taller species should be incorporated into small and dwarf kangaroo paws.

Some basic information on the breeding of kangaroo paws has been developed. The color and vigor of the foliage of *A. flavidus* was dominantly inherited in all hybrids. A hybrid of *A. rudus* x *A. flavidus* was the largest, reaching up to 5 feet tall, while *A. kalbarriensis* x *A. flavidus* formed the smallest with a densely flowered stem—only 8 to 12 inches tall, with short, broad leaves.

A. rufus x *A. flavidus* hybrids were the first to flower (as early as July), producing more flowers per plant as compared to the individual species. *A. manglesii* x *A. Flavidus* produced fewer flowers than *A. flavidus*, although the number exceeds typical wild *A. manglesii*.

Flower colors were variable among the hybrids, particularly the hybrid *A. manglesii* x *A. flavidus*. The crosses *A. preissii* x *A. flavidus* and *A. onycis* x *A. flavidus* have orange and orange-pink flowers, respectively.

Three cultivars with horticultural merit have been introduced to the industry from S.D. Hopper's breeding program. They are Dwarf Delight, Regal Claw, and Red Cross from crosses with *A. onycis*, *A. preissii*, and *A. rufus* as seed parents and *A. flavidus* as the pollen parent. At least two other major, ongoing, Australian breeding programs have begun to release additional cultivars.

Kangaroo paws have potential as a cut flower and pot plant for increased field or greenhouse production in cool US climates. The New Crops Program at the USDA-ARS, florist and Nursery Crops Laboratory in Beltsville, Maryland, will continue evaluations for hybrid kangaroo paws as cut flowers and pot plants. Basic work on temperature and photoperiod requirements for flowering is completed. Further trials to investigate growing media and nutritional requirements will be initiated in 1986 at Beltsville, Maryland, and with floriculture researchers at California Polytechnic State University, Colorado State University, and the University of Maryland.

"

Mark S. Roh and Roger H. Lawson are with the USDA-ARS, Florist and Nursery Crops Laboratory, Beltsville, Maryland. Robert G. Anderson is with the University of Kentucky, Department of Horticulture and Landscape Architecture, Lexington, Kentucky.

Acknowledgement: Plant materials used for the New Corps Research Program were supplied by Mervyn Turner, Bush Gems Garden Nursery, Monbulk, Victoria, Australia.

Culture notes

September 1989

by Meredith Shank

Greenhouse forcing snaps
Family: Scrophulariaceae
Genus, species: *Antirrhinum majus*

North American production of greenhouse forcing snaps is increasing. First introduced in 1938, F_1 hybrid forcing snaps peaked in popularity in the late 1950s, and in the last four or five years have seen a resurgence in popularity. Due to rising demand for cut flowers, better shipping techniques and economies of producing this cool crop, the demand for a high-quality local supply is creating a profitable price for the local grower. White, pink and yellow are the most popular colors year-round, with increased demand for red and bronze in the fall.

Germinating seed: Tips for success

Snapdragons are commercially propagated solely by seed. Use a commercial peat lite seedling mix or any media that is sterile, fine textured, with good drainage, aeration and water holding capabilities to germinate seed. Sow the seed thinly and in rows to minimize disease.

Optimum germination temperature is 68 to 70 F. Don't cover seed, but protect it from drying out with mist, a germination chamber or by covering with a plastic tent. Germination occurs seven to 14 days after sowing.

Condition the seedlings after removal from the germination area by placing flats on the bench in light to moderate shade. A reduced temperature of 60 F and good air movement prevent damping off. Supplemental lighting will benefit seedling development, especially in winter. At PanAmerican Seed we use 16-hour days with 700 footcandles, along with bottom heat.

Soil mix and spacing

Transplant when the first true leaves develop, about four weeks after sowing, to minimize disease and provide more uniform bloom quality. Many growers sow or purchase cut flower snaps from plugs. The crop is well suited to singulated seeding, and plug grown snaps give a uniform crop. Culture is the same, but transplanting occurs later, at five to six weeks, when the plugs lift easily from the tray.

Normally plants will flower up to two weeks earlier if grown on raised beds rather than ground beds, particularly during winter months. Ground beds can

Antirrhinum

also present more drainage problems than raised beds.

The soil in the beds must be well drained and have good aeration. The heavier the soil, the longer it will take the crop to bloom. Use a commercial peat lite mix or a mixture of peat soil, sterilized at 180 F for 30 minutes to prevent disease.

Most growers produce snaps as single stems, as it's possible to save three to four weeks' time over the pinched crop, plus considerable labor, and the crop finishes more uniformly. (Single stem culture allows three to four crops per year against two for pinched crop.) Don't grow pinched crops from June 15 to November 1. Spacing for the single stem crop should be 4 by 5 inches for a winter crop; use a 4- by 4-inch spacing in the summer. A pinched crop should receive a spacing of 7 by 8 inches. If a crop is pinched, prune it to four good breaks.

Snaps are light feeders

Snaps aren't heavy feeders; a low to medium nutrient level grows quality stems. A Spurway reading of 25 to 30 parts per million N, 5 to 8 ppm P, 25 to 30 ppm K and 150 to 200 ppm Ca is recommended. Fertilizer frequency and amount depends on soil structure and crop and weather conditions.

Avoid high nitrogen levels during winter to prevent soft bushy stems and grassy growth. Young plants respond better to a nitrate form of nitrogen than to ammonium forms. Stop fertilizing when first colors appear on the spikes. Avoid high soluble salts: Leaf chlorosis and short, weak stems will result. A media pH of 6.0 to 6.5 for soil and 5.5 to 5.8 for soilless mixes is best.

Snapdragons are divided into four flowering or response groups. Selection of forcing snap varieties is critical, depending on location. Growers must adjust scheduling for their own localities.

Response group	Approximate flowering dates	Comments
I	Dec. 1 to Feb. 20	Use a 50 F night temperature; sow seed Aug. 15 for Christmas flowers.
II	Nov. 1 to Dec. 15 Feb. 20 to May 1	Require longer time to flower than Group I varieties from Feb. 20-May 1.
III	April 10 to June 15 Sept. 15 to Nov. 1	Use a 60 F night temperature for quickest flowering.
IV	June 15 to Sept. 15	Quick flowering.

Antirrhinum

Grow quality snapdragons at 50 F night temperature. All scheduling and timing is based on this temperature. Daytime temperature should be 55 F on cloudy days and 60 to 65 F on bright sunny days. Condition seedlings to 60 F.

Clean glass or plastic is essential for quality flower production in periods of low natural light. In summer, a light shade may be necessary. Snaps are tall growing plants and require support. Commercially available plastic and metal mesh work very well. Normally three to five layers provide adequate support. If the flower spikes are allowed to tilt or lean, they will be unacceptable for sale.

Avoiding diseases

The most common snapdragon disease is Botrytis, which attacks when plants are starting to flower. This normally occurs during winter months. Pythium is a problem when excess soil moisture is present. Avoid over-watering, and use a well-drained soil with good aeration. Avoid drops and over-watering to control Rhizoctonia as well, and drench with Terraclor.

The most common insect pests are aphids, thrips, whitefly and—on occasion—mites. Screen greenhouse vents to prevent bees from entering and pollinating open flowers.

"

Meredith Shank is with PanAmerican Seed, West Chicago, Illinois.

Aphelandra

Culture notes

June 1985

by Teresa Aimone

Zebra plant
Family: Acanthaceae
Genus, species: *Aphelandra squarrosa*

The striking contrast between the green and white foliage and the bright yellow flower spikes make zebra plants popular. Plants are native to Brazil, and since the colorful yellow bracts are long-lasting, zebra plants work well in both interior landscapes and the home.

Aphelandra

Propagation

Zebra plants are best propagated from cuttings that have two fully developed leaves and an expanding terminal tip. Place the cuttings in a 1:1 peat/perlite medium, dipping the cuttings in rooting hormone prior to sticking the bottom 1 to 1½ inch of the cuttings into the medium. Maintain a soil temperature of 80 F (zebra plants respond well to bottom heat), and mist the cuttings for 30 seconds every 15 minutes from 9 a.m. to 4 p.m. until rooting occurs. The quality of the cuttings will also improve by either incorporating 14-14-14 Osmocote at eight pounds per cubic yard into the medium or watering with 20-20-20 at two pounds per 100 gallons twice a week.

Cultural requirements

Watering and feeding: Zebra plants require more water than many foliage plants, so allow just the soil surface to dry out between waterings. If, for some reason, the plants do wilt, water them frequently until they revive.

Feeding levels are related to light levels and growth rates, so the recommendations vary. When feeding is done, through, supply plants with 200 ppm nitrogen, 75 ppm phosphorus, and 150 ppm potassium. During the high light and rapid growth stage of the summer months, feed with every watering; winter feedings may vary from once every two weeks to once every two months. Carefully watch plant growth and adjust accordingly.

Humidity: Maintain 40 percent humidity levels; otherwise, leaf tips may turn brown.

Shading: Zebra plants require approximately 30 percent shade for best growth.

Temperature: Best growth response is seen when night temperatures do not drop below 60 F and soil temperature is between 70 and 80 F. Day temperatures should not go higher than 90 F, or physiological problems will result.

Aphelandra

Medium: Though aphelandras require lots of water, the medium should still have good drainage. A soilless mix will work well, but a 1:1:1 soil, peat, and sand mix is fine, too.

Flowering: Plants are usually sold just as the bracts at the bottom of the flower spike are beginning to open. And though uniform flowering is a bit difficult, the end result is well worth the effort, Flowering seems to be related to light—mainly light intensity and duration. Zebra plants are "photoaccumulative" and flower when light intensity is high or when lower intensities are received over longer periods of time. Experience has shown that best flowering occurs when plants are given 1,000 to 1,500 footcandles during the long summer days; during short winter days, zebra plants will flower when given 2,500 footcandles.

Physiological disorders

Crinkle leaf: With this condition, the leaves are crinkled, small, and the internodes are short. Numerous axillary buds will also appear. The best control is to follow prescribed temperature and light levels, avoiding high light and high temperatures.

Premature flowering: Premature flowering is associated with flower buds appearing on cuttings in the propagation area, the problem should go away.

Leaf drop: When the lower leaves begin to abscise, the grower should be certain to maintain proper moisture and humidity levels. Leaf drop can also be prevented by avoiding close spacings and excessive fertilizer levels.

Diseases and insects

Botrytis blight can be controlled on zebra plants with Chipco 26019 (Rovral). Pay close attention to plants during the winter months-especially the cuttings rooting in the misted propagation area. High humidity also may induce two leaf spot diseases, *Corynespora* and *Myrothecium* leaf spot. Both can be controlled with Daconil and Benlate. Corynespora leaf spot appears as brown or black, moist-looking spots on leaf edges, tips, and centers. Myrothecium has the same symptoms on the top of the leaves, and on the underside, black and white fungal fruiting bodies appear in rings inside the necrotic spots. Another way to help control these two types of leaf spots is to eliminate overhead watering.

Safe	Unsafe
Bacillus thuringiensis	Dursban EC
	Omite WP
Diazinon EC	Orthene WP
Dimethoate EC	Permethrin EC
Disyston EC	Resmethrin EC
Enstar	Vydate EC
Kelthane EC	Safer's Agro-Chem Insecticidal Soap
Metasystox-R EC	
Pentax WP	
Sevin WP	

Table 1. Phytotoxicity of insecticides and miticides on aphelandra. Table from *Foliage Digest*, January 1984. All pesticides were tested at recommended rates and intervals.

Phytophthora stem rot appears as black, soft lesions at the soil line. As the disease progresses, these lesions will move up the plant and cause eventual plant collapse. **Pythium root rot** is characterized with wilting and/or yellowing of the upper plant portions. Roots will be stunted, black, soft, and the outer portions of the root will be easily separated from the core. Control both Phytophthora and Pythium with Subdue.

Zebra plants are susceptible to most common greenhouse insects, unfortunately. **Mealybugs** can be controlled with chemicals such as Enstar 5E, Mavrik, and Bendiocarb; systemic insecticides tend to give better protection.

Use Metasystox-R, Disyston or Dimethoate. All these chemicals can be used to control scales on zebra plants, too.

For control of **mites**, use Kelthane, Pentac or Vendex. For **aphids and thrips**, insecticides commonly used to treat these insects can be applied to zebra plants. Check the label to see if the chemical is registered for use before application. The enclosed tables show pesticides which are either safe or unsafe to use on zebra plants.

"

Teresa Aimone, former editor of GrowerTalks *magazine, is with Sluis and Groot, Fort Wayne, Indiana.*

Astilbe

Culture notes

October 1990

by Jim Nau

Astilbe (False spirea)
Family: Saxifragaceae
Genus, species: *Astilbe* x *arendsii*

Astilbe x *arendsii* is a premier perennial plant with excellent value for either the potted plant market or for perennial plant sales in the spring. Astilbe flowers in late spring and early summer and does best in areas where it receives light shade in the afternoon. Plants get 24 to 36 inches tall and come in colors of red, pink white or lavender. Flowers are borne in plumes that measure from 4 to 10 inches long. Astilbes are hardy in USDA zones 4 to 8.

Propagation
Though astilbes can be propagated from seed, the boldest colors and all the hybrids come from divisions that are done in early spring or fall. Upon receipt of your roots, the crowns can be potted as is or separated using one plant per quart or gallon container. February and March planted divisions will be saleable by late May though flowering will not be profuse from these divisions, more flower color will be on next years plants.

For potted plant sales in flower during late winter and spring, bring the crowns (try to have three to four "eyes") in around September. Pot up into 6-inch pots and let the plants become established in the container. Chill the plants for 10 to 12 weeks at 35 to 40 F. They can be overwintered with your other perennials and brought in when needed. Once the pots have been brought

Astilbe

back into the greenhouse, place them out of direct light and allow to warm up gradually over a period of several days, then increase temperatures to no less than 58 F nights and allow 12 to 16 weeks to flower depending on the cultivar.

In general, the early and midseason varieties will flower about the same time, the late season varieties tend to take longer though they would work for sales two to four weeks after the start of the sales on the earlier flowering material. It is also suggested you stay away from taller varieties to avoid too vigorous a growth. A growth regulator is suggested for keeping the plants dwarf.

For those who would like to grow from seed (384,000 seeds per ounce) pretreat by placing the seed on moist peat moss at 70 F. Seed should be exposed to light and will germinate in 14 to 21 days. If the seed does not germinate, then give 40 F for three to four weeks. This will help to break dormancy within particularly hard to germinate seed. However, when the seed is fresh, 70 F has been all we needed to do to get 70 plus percent germination rates. If you follow the procedure above, once removed from the 40 F cooler, the seed will germinate in 14 to 21 days at 70 F. January sowing will not flower the same season from seed.

Astilbe Sentisifloea

Cultivars

In varieties, all seeded varieties are mixtures of the species. While the plants perform well the flower color is muted at best and dull in its overall appearance. As for the astilbe varieties from division, the following cultivars are suggested: Fanal (a deep red to 22 inches), Rheinland (a pink to 24 inches) and Deutschland (a pure white to 24 inches). These as well as other varieties on the market are classified by being from early to late season performers; check with your favorite supplier in regards to the seasonality of the cultivars you are interested in growing.

In related material, *Astilbe chinensis* var. pumila is an excellent variety with lavender pink flowers that prefers a well-drained though moist soil in the garden. This dwarf variety (10 inches) is recommended for use in rock gardens and as a border or edging plant in the perennial garden. It also makes a sharp 4-inch pot. *A. chinensis* var. pumila flowers in August in the Midwest garden, making it one of the latest to flower.

99

Jim Nau is trials and product development manager for seed at Ball Seed Co., West Chicago, Illinois.

Azalea, see Rhododendron

Bedding Plants

Sowings

December 1987 to November 1988

by Nancy Bogart

JANUARY

This month, start getting ready for May sales of hanging baskets, bedding flats and spring pot crops.

Hanging baskets

Sow seed of petunia, impatiens, portulaca and thunbergia for early-to mid May sales. All of these crops require a constant 70 F soil temperature for germination. The petunias and portulaca should be sown on the soil surface and not covered with media. Cover both the impatiens and thunbergia seed with ¼ inch of soil to maintain uniform moisture. All seed should be well germinated and ready for transplanting in 3 to 4 weeks. Transplant to 2¼-inch cells and maintain 62 to 65 F to get plants established. Plants will be ready to go to their final containers approximately 3 to 4 weeks after the initial transplanting. For 8-inch hanging baskets plant three seedlings per basket; for 10-inch baskets use five seedlings. Continue to grow all of these on at 62 to 65 F except for the petunias. Once established in the final container drop the temperature to 55 F for compact, bushy baskets.

Petunia varieties that perform well in baskets are the Madness series-all colors make spectacular baskets. They stay compact and continue to produce a mass of flowers all summer. The Cascade varieties are also suited for hanging baskets. Almost any variety of impatiens will perform well in a hanging basket but the Showstopper series and Blitz series are exceptional. Try the Susie series of Thunbergia, available in three separate colors and a mix. Remember to train the seedlings up the basket wires to produce a full, bushy and upright basket. The Sundial series or Wildfire Mix portulaca are suited for baskets.

Flat bedding

Vinca seed should be sown in early to mid-January for mid-May sales. The seed should be covered with germination media to provide both darkness and uniform soil moisture. Proper germination temperatures for vinca are critical—70 to 75 F soil temperature must be maintained until all seed has germinated. With proper temperatures, seed should germinate in about 15 days. Once the seed has germinated maintain 65 to 70 F temperatures. Watch seed flats closely for damping off disease. Seedlings should be ready for transplanting 4 weeks after sowing. For the first 3 weeks after transplanting be careful not to overwater or let temperatures drop below 60 to 65 F.

Bedding Plants

Begonias should be sown in mid-January for early to mid-May sales. Begonia seed should be sown on the surface of the media and left uncovered. A minimum temperature of 70 F must be maintained for germination. Generally it will take a good 2 to 3 weeks for seed to germinate. Grow on in the seed flats for another 6 to 8 weeks before transplanting. A very light feed of 100 to 150 ppm of a balanced fertilizer seems to make them jump in the seed flat, sometimes actually reducing crop time by 5 to 7 days.

Potted bedding

Seed geraniums are so uniform and easy to grow that everyone should try them. Sow the seed in mid-January to have blooming 4-inch pots in early May. Seed should be sown thinly in flats and covered with ¼ inch of media. Germination temperatures are critical, maintain 77 F. Seeds germinate in 7 to 10 days. Not all varieties germinate at the same rate. Transplant to flats or direct to 4-inch pots when the first true leaf appears (approximately 3 weeks after sowing). Grow on at 62 to 65 F. Plants should be treated with 1500 ppm of Cycocel when there are three leaves the size of a 50-cent piece, or about 3 weeks after transplanting. Another application can be given if needed 7 to 10 days later. Be sure to apply Cycocel as a fine mist and not to run off to avoid yellowing and burning of the foliage. Once flower buds are visible, 60 to 62 F temperature will help to harden off the plants.

The tomato variety Patio makes a tempting 8-inch pot when seeds are sown in mid-January. Plants will have fruit that is nearly ripe for early to mid-May sales. Sow seed and cover lightly with germination media. Seeds will germinate in 7 to 10 days with 72 F soil temperatures. Transplant seedlings to 2¼-inch cells or peat pots as soon as they are large enough to handle. Grow on at 62 F until plants have filled the cells. Transplant to an 8-inch pot and continue to grow at 62 F until plants are 6 to 8 inches tall. Once they have reached 6 to 8 inches, place a 2- to 3-foot stake in the pot for later support, and drop temperatures to 55 to 60 F to produce a very thick, sturdy stem. Grow on at 55 to 60 F until sale, and be sure to periodically tie plants to the stake to support the full-grown plants.

FEBRUARY

Ageratum

Sow seed mid-February for blooming flats May 1 or 4 inches for late May. Seed should be sown on well-drained germination media and exposed to light. Seed will germinate in eight days in 70-75 F temperatures are maintained. Four weeks after sowing, seedlings should be ready for transplanting to 2¼-inch cells. Be sure to transplant the seedlings up to the first true leaf to obtain a uniform, compact, finished flat. B-Nine also will help to produce uniform flats, as well as enhance deep green leaf color. B-Nine can be used at 1,500 ppm when seedlings have filled out the 2¼-inch cells and another application can be given one week later. Watch for Botrytis on the terminals of the seedlings the first three weeks after transplanting. Spray with an appropriate fungicide at first sign of disease. Maintain at least 60 F after transplanting; for a nice compact 4 inches, hold seedlings in the 2¼-inch cells until a flower bud has formed, then shift to the final 4-inch container. For a darker lavender-blue color try Royal Delft. Madison is a unique periwinkle blue and Blue Hawaii is very early and uniform. All varieties are 6 inches at maturity and do well in flats and pots.

Dusty Miller

The Silver Dust variety has silvery white foliage and tolerates pruning to produce short bushy plants. Seed of this variety should be sown at mid-February for flats in early May; in early February for 4-inch pots in mid-May. Soil temperatures must be maintained at 75 F and seed should be exposed to light for this particular variety to germinate. Germination should occur in 10 to 15 days. Transplant to packs or 4-inch pots four to five weeks after sowing. Maintain 62 to 65 F after transplanting, until several weeks before sales, then drop temperatures to 55 F to keep plants compact, bushy, and harden off before going outdoors.

Lobelia

This plant offers a different shade of brilliant blue than most other annuals. Seed should be sown in rows in mid-February for salable flats in mid-May. Do not cover seed with germination media. Germination will occur in 20 days if 70 to 75 F soil temperatures are maintained. Five to six weeks after sowing, transplant in clumps to cell packs. Grow on for the next few weeks at 60 F and then drop the temperature to as low as 45 to 50 F to maintain very compact flats. The variety Blue Moon is up to a week earlier than other varieties and has green foliage with dark blue flowers. It performs nicely in cell packs as well as 4-inch pots. Crystal Palace is also a brilliant dark blue but has bronze green foliage and is best used in flats. For unusual hanging baskets, try the Fountain series. Lilac Fountain is especially nice.

Petunias

Varieties of double petunias should be sown in early February to produce flowering packs in early to mid-May. Single petunia varieties can be sown two weeks later to produce colorful flats at the same time. Seed can be mixed with a small amount of sugar to make sowing easier. Sow seed on germination media, but do not cover the seed. Maintain 70 F for uniform germination in 10 to 12 days. Transplant seedlings three to four weeks after sowing and grow on at 60 F for the next two weeks until seedlings are established, then drop the temperature to 50 F and grow on to finish. Cooler temperatures will produce very compact plants. The new floribunda (multiflora) petunias, such as Madness, Carpet, Celebrity, Primetime or Polo, offer the most compact plants that are covered with masses of blooms.

Portulaca

Sow seed in late February for early May sales. Seed can be sown directly into cell packs or sown in flats and transplanted in clumps. Soil temperature is critical; maintain 70 F. Germination will occur in 10 days; once the seed has germinated try to run slightly on the dry side. Portulaca is very susceptible to damping off disease in the seed flats. Treat with a fungicide if necessary. Once seedlings are transplanted (about four to five weeks after sowing) grow on at 65 F. The F_1 Sundial series, available in separate colors, is fully double and opens earlier and stays open longer than other varieties. Use only two or three seed per cell to set the best show from Sundial.

Bedding Plants

MARCH

Alyssum

Seed sown in mid-March will produce colorful flats for early to mid-May sales. Seed can be directly sown into packs or transplanted in clumps. Seed should be left uncovered and exposed to light for good germination. Soil temperatures of 70 F will germinate seed in eight days. Watch carefully for damping-off disease in the seed flats. Transplant seedlings in clumps of four or more seedlings about three weeks after sowing. Grow on at 60 F for two weeks until seedlings are established, then drop temperatures to as low as 45 F. Grown cool, alyssum stays very compact. Wonderland is the mainstay alyssum, but for a twist try one of the pastel mixes such as Easter Bonnet, Pastel Carpet or Golfy. Also, Snow Crystals has shown better heat tolerance than other alyssum in the South.

Dahlias

For blooming packs in mid-May sow seed in mid-March. Seed should be covered with 1/8 inch of soil and the media should be kept moist. Seed will germinate in 10 days if soil temperatures are kept at 70 F. Seedlings will be ready for transplanting 13 to 21 days after sowing. Transplant seedlings up to the cotyledons. Drop the temperature to 60 to 65 F after transplanting. For 4-inch pots start approximately three weeks earlier and feed heavily so plants will branch freely. Figaro is the most uniform seed dahlia and shows bright large flowers. For a dark-leafed variety, try Diablo as a pinched plant.

Impatiens

Seed sown in early to mid-March will produce flowering packs for mid-May sales. Seed should be sown thinly on a well-drained medium. Impatiens do not require darkness to germinate, but covering the seed lightly with germination media helps to prevent moisture loss around the seed. If a mist system is available, there is no need to cover the seed. Maintain uniform temperatures of 70 F and seed will germinate in 18 days. After germination grow seedlings on at 60 to 62 F. Four weeks after sowing transplant into packs. Several applications of a balanced fertilizer after transplanting will help to get seedlings established and off to a good start. Thereafter clear water or feed lightly. Too much feed through the entire crop will cause excessive vegetative growth and few flowers.

Marigolds

Sow seed of French and African type marigolds mid-March for blooms in early to mid-May. All seed should be sown and covered with germination media. Maintain 70 F soil temperature for germination. Immediately after seed has germinated lower temperatures to 65 F. Seed should germinate in seven days. Be ready to transplant seedlings to packs 10 to 14 days after sowing to avoid stretching in the flats. Don't forget African type marigolds are daylength sensitive so you will need to provide short days for 14 days after sowing to insure uniform flowering. You can create short days by covering seed flats with black plastic or an inverted flat without holes at 5 p.m., and removing the cover at 8 a.m. Once seedlings are established after transplanting they can be grown quite cool to control height. Drop temperatures gradually to 50 F, or even 45 F, if you want to hold the crop. Offer an assortment of sizes and colors.

Bedding Plants

Salvia

Sow seed in early March and sell in packs prior to flowering in early to mid-May. The only variety you may want to sell with color is Red Hot Sally, plants hold well. Seed should be sown and left uncovered as light enhances germination. Maintain 70 F soil temperatures and seed will germinate in 12 to 15 days. Be sure to supply adequate moisture either by using a mist system or covering flats with clear plastic. Transplant to cell packs three to four weeks after sowing. Salvia can be grown cool (55 to 60 F) to produce compact flats. For colored series, select Sizzlers for early, uniform pack performance; Empires for a medium-sized landscape product or Cleopatras if you need a variety for pot production with outstanding field performance.

APRIL

April is the month to finish sowings for May bedding plant sales and to start sowings for 4-inch annuals.

Nicotiana

Nicotiana should be sold green in the pack, so allow about seven weeks from sowing to sale. Seed sown in early to mid-April will produce salable packs in mid- to late May. Seed must be exposed to light to germinate, so *do not* cover with media. Soil temperatures of 70 F are also critical for proper germination. Seed will germinate in 15 days if these two requirements are met. Seedlings can be transplanted as soon as they can be handled, approximately three weeks after sowing. Grow on at 60 F or cooler if necessary to keep them compact. Nicotiana also responds to B-Nine applications. The Nicki series offers a Red, White, Pink, Rose, Yellow and a Mix. It makes an outstanding self-branching bedding plant. The Domino series is more compact and comes in Pink with white eye, Purple, Red, White and a Mix. The earliest nicotiana variety is the Starship series.

Peppers

Gardeners should not set out peppers until all chance of frost is nonexistent. Sowing seed in early April will produce quality plants in packs for mid- to late May sales. Soil temperatures of 72 F are critical for germination. With proper temperatures seed will germinate in 10 days. Transplant to packs 24 to 30 days after sowing. Grow on at 62 F nights and 70 F days for May sales. Bell peppers are by far the most popular types. For a really hot chili pepper try Super Chili. Mexi Bell offers a mildly hot flavor. Whopper Improved bell pepper offers thick-walled fruit, while Mr. Bell 7 is an early (60 days) bell pepper with blocky fruit. For colorful fruit try Orobelle with golden yellow fruit or Purple Bell for eggplant-colored fruit.

Tomatoes

For salable plants in packs and 4-inch pots, sow April 1 for mid- to late May sales. Seed should be covered lightly with germination media and soil temperatures of 72 F maintained for proper germination. Seed will germinate in seven to 10 days and should be transplanted within a week after germination. Transplant seedlings deep and grow on at 65 F. Once plants reach a salable height begin to reduce the temperature gradually to harden-off plants and maintain a good height. There are a number of outstanding tomato varieties for bedding plant sales, consult your seed salesmen.

Bedding Plants

Ornamental peppers

These make a nice fall mass-planting annual to replace old summer annuals. They are also great as small flowering pot plants. Grow and sell in 4-inch pots. For sales in early to mid-September sow seed in mid- to late April. Similar to bell peppers, ornamental peppers must have 72 F soil temperatures to germinate. Cover seed lightly with germination media. Seed will germinate in 12 days. Transplant to 2¼-inch cells or directly to 4-inch containers as soon as seedlings can be handled. Grow ornamental peppers on at 65 F. Firework is a popular variety that has an abundance of cone-shaped fruit that sets uniformly above the foliage, fruit changes from cream colored to a brilliant red. Holiday Cheer has round fruit that goes through several color changes before turning its final red color. Treasure Chest is early and shows cone-shaped fruit.

MAY

Flowering annuals

Keep garden centers and landscapers supplied with fresh annuals throughout the summer by sowing in May. The big sellers, impatiens, marigolds and petunias, can be sown this month for July sales in 4-inch pots.

Impatiens: Sow seed May 1 for mid- to late July sales. Be sure to cover seed with media and to supply adequate moisture for germination or use a mist system. Maintain 70 F temperatures and seed will germinate in 18 days. Transplant seedlings directly to 4-inch pots as soon as seedlings can be handled—approximately three weeks after sowing. Where possible drop the temperature to 60 F for growing on. Plants will require some shade during the months of June and July.

Marigolds: Sow marigold seed in early May for early to mid-July sales. Seed germinates in five to seven days if 70 F temperatures are maintained. Within five to seven days of germination transplant seedlings directly into 4-inch containers.

Petunias: Sow seed in early May for mid-July sales. Germination will occur in 10 days for most varieties if soil temperatures are maintained at 70 F. Transplant seedlings as soon as they can be handled. Grow on as close to 60 F as possible.

JUNE

Coleus

Sow in late June to be ready for sales in early to mid-September. Grown in a 4-inch pot, coleus makes a colorful foliage plant for fall sales in the North, or a shady annual in the South. Sow in rows and maintain constant 65 to 95 F soil temperatures. Seed must be exposed to light, so do not cover with germination media. Germination will occur in 10 days. Transplant about 5 weeks after sowing. Grow on at 60 F nights. Flower buds should be pinched as soon as they begin to form. The Wizard series works well in 4-inch pots. It has large heart-shaped leaves and stays fairly compact.

SEPTEMBER

Pansies for the West

Early September sowings of pansies along the Pacific Coast will produce salable 4-inch pots in December and January. Cover seed lightly and maintain

60 to 65 F soil temperatures. F_1 hybrids will give the best performance. For a profusion of medium-sized flowers on an exceptional garden performer, try the Universal series. Universals are available in a wide range of blotched and bicolored and clear colors as well as a Mix. For clear faced straight colors and medium sized flowers, try the Crystal Bowl series. For a large flowered series with mostly blotched patterns, try the Majestic Giants series in separate colors or a mix. The Roc series, blotched with large flowers, are known for the shippability and weather resistance their sturdy stems give them. For a large flower in a clear faced Mix or separate colors, try the Crown series. For a large-flowered retail sales area winner, try Imperial Pink Shades or Imperial Silver Princess, both pastels.

OCTOBER

Asparagus sprengeri: Sow seed at the end of October for 4-inch pots in early May. Seed should be sown in rows in a moist sterile media. Cover the seed with approximately 1/8 inch of media. Keep soil temperatures at 75 F nights and 85 F days until germination is complete. The seed does not always germinate uniformly, allow 21 to 45 days for complete germination. Be sure to provide plenty of moisture for seed to germinate. When seedlings are between 1 and 3 inches tall, transplant them into 4-inch pots. Three seedlings should be placed in a loose clump in each pot. Grow on at 60 to 62 F, being sure to keep them moist and well fed.

Begonias (tuberous): Late October sowings of tuberous begonias make gorgeous 4- and 8-inch hanging baskets for early to late May sales. Seed should be sown on a very light, well-drained media. Since begonia seed is so small, some growers prefer to mix the seed with sugar or Sweet'N'Low to make handling easier. The seed can be sown either in rows or broadcast over the seed flat. Seed should be exposed to light so do not cover with media. Adequate moisture is a must for begonias to germinate. A mist system is ideal. Use a clear plastic covering if a mist system is not available. Soil temperature should be kept at 70 F, no less, until germination is complete. Seed will germinate in 15 to 30 days. After germination is complete, reduce the amount of mist slightly. Once seedlings are visible, night lighting is necessary to prevent tubers from forming. Plants tend to stall out and stop vegetative growth if a tuber forms early. To supply the proper daylength, turn on night lights from 10 p.m. to 2 a.m., or for four hours after dark. Lighting should supply at least 10 footcandles of light and should be given from October to March. (We have found that we get a significantly better crop if we continue to light through March.)

Continue warm temperatures, even after germination—maintain 65 F. Begonia seedlings spend a considerable amount of time in the seed flats, sometimes as much as two to three months before transplanting. Once seedlings are large enough to be watered, without being washed to the sides of the flats, apply a diluted strength balanced fertilizer at 100 to 150 ppm. After that feeding seedlings will visibly jump in the seed flats.

Once the seedlings are large enough to handle, transplant to 2¼-inch cells. Continue with warm temperatures and night lighting. Approximately six to eight weeks later they will be ready to be shifted to the final containers. Put one 2¼-inch per 4-inch and three 2¼-inch per 8-inch hanging basket. Grow on at 62 to 65 F, no less.

Several varieties offer magnificent colors. The Non-Stop Series are all F_1 hybrids and offer a nine-color selection as well as a mix. The variety Clips is about three weeks earlier than Non-Stop and very compact. For a more upright variety try the new Memory Mixture, and for hanging baskets try the Musicals. Non-Stop Ornament has bronze leaves.

DECEMBER

The bedding plant season will be fast approaching. There are several annuals that can be sown in December to start sales off in March and April.

Pansies

Seed sown in early to mid-December will produce flowering 4-inch and flat material for sales from late March through early May. If pansies are grown and kept cool they have excellent shelf life, extending your shipping and sales period to as long as six weeks. Seed should be sown in rows and covered lightly with germination media. At 65 to 70 F soil temperatures, seed will germinate in 10 days. Approximately three to four weeks after sowing seedlings will be ready to be transplanted to 4 pots or cell packs. Allow seedlings to become established by maintaining 60 F nights for the first 10 to 14 days after transplanting. Once seedlings are established, begin to drop the night temperature to 45 to 50 F. Grow on at cooler temperatures to produce sturdy, compact plants.

Calendula

Sow seed of this crop in late December for mid-February sales in the South. Calendulas thrive in cooler temperatures and are popular in the South. Seed should be sown thinly and covered with media to block out light. Maintain 70 F soil temperatures; seed will germinate in 10 days. Once seed has germinated, drop temperatures to 60 to 65 F nights. Transplant into cell packs or 4-inch pots when seedlings are large enough to handle. Calendulas are fairly large and vigorous seedlings; do not put more than 48 plants per flat. Grow on at 45 to 50 F to produce a nice compact plant.

The Bon Bon Series is a very dwarf variety with 2½- to 3-inch blooms. These can bloom 14 days ahead of the other varieties. For a more heat-tolerant calendula try the Coronet Series.

Browallia baskets

Seed should be sown in mid-December to produce 10-inch hanging baskets for early to mid-May sales. Sow seed on the surface of the germination media; do not cover with media. Maintain an even 70 F soil temperature for uniform germination, then lower temperatures to 65 F to grow on in seed flats. Seed will germinate in 15 days. Transplant to 2¼-inch cells approximately four weeks after sowing. Continue 60 to 65 F for the remainder of the crop. Shift seedlings to final 10-inch hanging basket as soon as plants fill the 2¼-inch cell. Three to five plants should be used in a 10-inch hanging basket. Keep plants pinched to produce bushy, well-branched plants. Stop pinching six weeks before desired flowering. It is advisable to harden plants off at 60 F for a week or so before moving outdoors.

Several varieties make beautiful baskets for a shady location. Blue Bells improved has 1½-inch blooms that are lavender-blue in color. Silver Bells is white; Sky Bells is a powder-blue and Jingle Bells is a mix of colors.

"

Nancy Bogart was formerly with DuPage Horticultural School.

Begonia, Fibrous

Culture notes

December 1990

by Jim Nau

Fibrous begonias
Family: Begoniaceae
Genus, species: *Begonia semperflorens cultorum*

Begonias are one of the best all around varieties to use in ornamental horticulture. They are one of only a handful of plants that maintain a profusion of blooms from the time they first bud up in June until the frost takes them in the fall. Well known as one of the most versatile landscape plants, begonias seldom need trimming to keep their shape and do not require dead-heading to lose the old blooms. Begonias work best in full sun or semi-shaded areas when planted 10 inches apart. Plants will fill in slowly but will remain in bloom all summer.

Propagation

Fibrous begonias have the smallest seed of any of the commercial crops sold on the U.S. trade today. You can expect approximately 2,000,000 seeds to the ounce, with most seed companies selling 1,000 seeds as the smallest quantity they sell.

In general, sow your seed on a sterile, peat lite media and water the seeds in. Do not cover with media, but allow contact between the media and the seed itself. If you do not have a mist system, cover flats with a pane of glass and remove immediately upon germination—otherwise seedlings will stretch.

In regards to germination temperatures, the sources vary as to the recommended way to germinate your seeds. Specifically, if you are using a sweat or germination chamber, the suggested temperatures for germination are 82 to 85 F for five to seven days; then temperatures should be dropped to 78 F for the remainder of germination. If you are germinating within a greenhouse environment, however, the germination temperatures should be 70 to 72 F. You will see a 2 to 3 percent decrease in germination compared to the sweat box method. While you can germinate in a greenhouse using 82 F, take care not to let the seeds dry out or germination will be erratic and growth uneven. Germination takes 14 to 21 days depending on the variety.

Because of their extremely fine root system, start dilute liquid feedings as soon as seedlings emerge to prevent post-germination stall. Once seedlings have emerged, move the sowing flat to a bright area of the greenhouse or place under fluorescent lights. The lights improve plant quality and decrease stretching—especially under the short days of winter. We leave them under

Begonia

Fibrous begonias offer a wide variety of colors for the landscape market.

lights until we are ready to transplant, which usually takes place after about 40 to 50 days after sowing.

Growing

When you are ready to transplant, use only one plant per 3-, 4- or 4½-inch pot when growing any of the standard commercial varieties from seed. On pot sizes 5 inches and larger, use one plant per pot if you are using varieties like Encore or Party. On varieties like Varsity, Olympia or Prelude, it's best to use two plants per pot to get the plants to fill in faster.

When transplanting, you can transplant directly from the sowing or plug flat to the final pack or pot. On pots, however, if the plants are small, you should transplant seedlings into cell packs and then into the final container. By doing so, you can control growth more readily.

When grown at 60 F night temperatures, green packs of 32 cells per flat take 14 to 15 weeks; flowering packs require up to 16 weeks to finish. Four- to 4½-inch pots take up to 18 weeks to finish. Keep in mind that begonias are one of the bedding plants that flowers before its roots are established enough to be transplanted to the garden. The plants will be in flower (though usually not profuse) several weeks before the roots have firmly anchored themselves.

When outside night-time temperatures are in the fifties, you can move begonias outside to cold frames to harden off plants prior to sale. This cooler air will help to tone up the foliage and make it thicker—bringing out the rich colors in either dark- or green-leafed varieties.

Varieties

Dwarf varieties: At 8 inches tall, Prelude is the dwarfest variety on the trade. Available in five colors and a mixture, Prelude, along with Varsity, are the earliest flowering varieties. The plants perform best in borders and containers.

Medium height varieties: Medium varieties are between 8 and 10 inches tall (and often-times 10 inches tall by late season) and are characterized by a large number of varieties. Some excellent choices include Pizzazz, a Park Seed Exclusive variety with four colors and a mixture; Varsity, characterized by medium-sized blooms on plants available in four colors plus a mixture; and Olympia, a series with five flower colors that has a rounded appearance all summer long.

In bronze-leafed varieties, Cocktail is still the leading variety among growers for a well-branched, dark-leaf companion to the above mentioned varieties. Several newer selections, Rio and Roxy, are proving their merit in U.S. gardens. Both varieties are available in four separate colors plus a mixture and have similar performance in habit and height in the garden.

Tall height varieties: Encore and Party are the two tallest varieties sold on the U.S. trade at this time. Both varieties grow from 14 to 18 inches tall in the garden and are available in both green- and dark-leafed varieties. One newcomer to this group is the All Round series. This series is available in three colors: Rose, a deep rose with bronze foliage, and White and Pink, which are both green-leafed varieties. All Round will be introduced into commercial catalogs in 1991.

99

Jim Nau is trials and product development manager for seed at Ball Seed Co., West Chicago, Illinois.

Begonia, Tuberous

Culture notes

May 1987

by Ron Adams

Hiemalis begonias
Family: Begoniaceae
Genus, species: *Begonia socotrana* x *Begonia tuberhybrida*

Hiemalis begonia is a crop with a wide range of showy flowers on contrasting dark foliage. They're the result of a cross between winter-flowering, bulbous species *Begonia socotrana* and the summer-flowering *Begonia tuberhybrida*. Even though the result of this cross has been known since 1883, its current success began when Otto Rieger from Germany introduced new varieties in 1955 that were both more floriferous and resistant to mildew. This group was called Rieger begonias. Several varieties were developed from these crosses and promoted throughout the United States by licensee, Mikkelsen's Inc., Ashtubula, Ohio.

Begonia

Today, there are in excess of 3 million Hiemalis begonia cuttings grown in a wide range of pot sizes.

Plants grown from multiple-stem cuttings do not require staking. They produce high-quality foliage resistant to powdery mildew and single, semi-double, and double flowers. Hanging baskets are generally produced from tip cuttings.

Propagation

Most Hiemalis begonias are patented varieties and are produced under license agreements. Cuttings are available as multi-stem, single-stem, and tip cuttings.

Long-day treatment and temperature

Hiemalis begonias should be grown at 70 to 72 F night temperatures during the first one to four weeks to encourage vegetative growth. Long days consist of 14-hour days with daylength extended in crops grown from September to March. Ideally, you should use at least 20 to 50 footcandles of light during long days to maintain vegetative growth.

A schedule for extended daylength would look like: March and September, extend daylength three hours a day; October and February, extend daylength four hours; November and January, extend daylength five hours; and December, extend daylength six hours.

Hiemalis begonias are sensitive to excessive high light levels, which cause sun scalding (marginal dessication and burning), and vegetative hardening that reduces growth. Ideal light levels vary by temperature: At temperatures of

Begonia

Spring/summer growing schedule*

Starter plant	Pot size	1st pinch	2nd pinch	Begin short-day treatment	Finish
Single-stem	4"	week 1-2	—	week 2-3	week 8-12
Multi-stem	4"	—	—	week 1-2	week 8-12
Single-stem/multi-stem	6"	week 2-3	—	week 4-6	week 10-16
Multi-stem	8"-10" basket	week 2-3	—	week 4-6	week 10-16
Multi-stem	8"-10" basket	week 2-3	week 5-7	week 7-10	week 14-20
Tip cutting	4"	week 1-2	—	week 2-3	week 8-10
Tip cutting	8"-10" basket	week 2-3	week 5-6	week 7-8	week 12-14

*Begin crop with long-day treatment.

65 F or below, maintain 3,000 footcandles; at temperatures of 70 to 80 F, maintain 2,000 footcandles; and at temperatures of 80 F or above, maintain 1,500 footcandles.

Short-day treatment and temperature

Short-day treatment should begin when the desired number of shoots have emerged and there is sufficient new growth. At this point, lower night air temperatures to 62 to 65 F.

Hiemalis begonias are only slightly photoperiodic. By giving a daylength of no greater than 12 hours, vegetative growth is slowed and flowers are more uniformly developed. Four weeks of short days are required to fulfill the short-day treatment. When using black cloth during periods of high night temperatures, it's best to pull the cloth from 7 p.m. to 9:30 a.m.

Growth regulator/pinching

Cycocel and A-Rest have proven to be useful in obtaining compact growth when needed. Cycocel at 1,500 ppm or A-Rest at 25 ppm should be applied 21 to 28 days after the start of short days.

Pinching is also effective for controlling growth, particularly when one or two shoots show dominance. Pinching can also be used to shape plants during long days. (Oglevee does not recommend pinching single-stem Spectrum series plants grown in 4-inch pots.)

Timing*

Pot size	Multi-stem cuttings per pot	Single-stem cuttings per pot	Tip cuttings per pot	Weeks long days	Final spacing (inches)	Weeks to finish
4"	1	1	—	1-2	6 x 6	8-12
6"	1	2	—	4-6	11 x 11	10-16
8" basket	3	—	—	4-6	14 x 14	10-16
10" basket	3	—	—	4-6	18 x 18	10-16
8" basket (double pinch)	1	—	—	6-7	14 x 14	14-20
10" basket (double pinch)	1	—	—	6-7	18 x 18	14-20
4"	—	—	1	1-2	6 x 6	8-10
8" basket (double pinch)	—	—	3	3-4	14 x 14	12-14
10" basket (double pinch)	—	—	4	3-4	18 x 18	12-14

*Timing schedule for plants started at 70 F and finished at 62 F. Four weeks is minimum short-day treatment.

Media and nutrition

Hiemalis begonias like a well-drained, well-structured media with at least 50 percent peat moss. Other media components may be perlite, vermiculite, composted pine bark, and mineral soil. Soil pH should be 5.0 to 5.5 for a soilless media, and 5.5 to 6.0 for a mineral soil-based media.

A complete fertilizer (including trace elements) should be used. A peatlite formulation such as 15-16-17 at 200 ppm nitrogen is recommended. Begin fertilizing when roots are established to the end of the container. To keep excess soluble salts from building up and injuring sensitive roots, leach pots after every third feeding. It's best to use a fertilizer low in ammonium nitrogen. Add extra calcium and magnesium to the fertilizer formulation, since most soluble fertilizers do not contain sufficient levels to maintain adequate media levels throughout the life of the crop.

CO_2

CO_2 benefits plants produced under low light conditions. CO_2 will increase photosynthetic activity at levels of 1,000 to 1,500 ppm. As a rule, greenhouse temperatures can be increased during the day by 5 to 10 F at higher CO_2 levels.

Insects

Cyclamen mites, aphids, thrips, whiteflies, mealy bugs, and caterpillars have been known to infest Hiemalis begonias. It's best to check with local cooperative extension agents for regional advise on pest control.

Foliar nematodes

Foliar nematodes (eel worms) are wormlike parasites of plants that infest stems and leaves. They cause small, yellowish spots that turn the leaf brownish black or on some cultivars, purple to red. The infected area takes on a water-soaked appearance on the underside of the leaf. As the first-line defense in nematode control, maintain strict sanitary conditions. It has been reported that Vydate has controlled foliar nematodes.

Diseases

Powdery mildew is common to Hiemalis begonias. It's easily recognized as a white fungus growing on the leaf surface. Powdery mildew may be controlled by using vaporizing sulfur burners or by spraying Karathane WP at 6 ounces per 100 gallons. Karathane will cause petal burn when sprayed on open flowers. Bayleton 25% WP has also been reported effective in controlling powdery mildew at a rate as low as 2 ounces per 100 gallons.

Xanthomonas begoniae is referred to as oil spot sickness or bacterial blight. The oil spot is identified as small, round spots evident on the underside of the leaf. They are usually noticeable as discolored spots that turn from yellow to brown. *Xanthomonas begoniae* affects stems as well as leaves and is intensified by high temperatures and high humidity. Control Xanthomonas by: (1) Securing clean stock from propagators using culture-indexed stock; (2) Not watering overhead. Use a drip system; (3) Complete all watering by noon; (4) Lower relative humidity by heating and ventilating; and (5) Removing and destroying all infected plants.

Botrytis typically develops in situations with high humidity and poor air circulation or by keeping plants' foliage wet for four to six hours continually. Reported controls are Chipco 26019, Daconil 2787, Exotherm Termil, Benlate, and Ornalin.

Because of the high disease potential in high humidity, this crop should be grown at a 60 to 65 percent relative humidity.

"

Ron Adams is technical services manager, Ball Seed, West Chicago, Illinois.

Brachycome

Culture notes

May 1986

by Teresa Aimone

Brachycome (Swan River Daisy)
Family: Compositae
Genus, species: *Brachycome iberidifolia*

Brachycome, commonly called Swan River Daisy, is another relatively new flowering pot plant. Uses for brachycome include: hanging baskets for spring sales in full sun locations; combination planters; outdoor plantings such as annual flower beds and borders; patio planters and urns, and mass plantings in parks, shopping centers, etc., where full sun is present. It also looks to be a great item for summer sales in 4-inch and larger size containers.

Medium
Brachycome best likes a porous, well-drained medium. An ideal mix that would provide good drainage yet sufficient water retention would be a combination of 50 percent peat moss and 50 percent aggregate (perlite, polystyrene beads, etc.). If you use an artificial mix, you may wish to try growing brachycome in this.

pH
For optimum growth, keep the pH regulated to around 6.0 to 6.5. If the pH gets any higher (becomes more alkaline), the leaves may turn yellow.

Temperature
Brachycome can be grown cool; recommendations suggest maintaining a night temperature of 60 to 62 F.

Brachycome

Strive to hold daytime temperatures below 85 F. If temperatures rise above this point, plants may have a reduced number of flowers and weak vegetative growth.

If the crop is coming in ahead of schedule and flower buds are visible, flowering can be delayed by reducing the temperature to as low as 45 F without damaging the plant.

Since brachycome is in the same family as chrysanthemum, it will respond to short-day treatment.

And, if plants are going to be grown during dark periods with short daylengths, added benefits will be achieved if CO_2 enrichment is provided. Supplement the greenhouse environment with 1,000 to 1,500 ppm CO_2 to encourage faster growth.

Light

High light conditions should be maintained in the greenhouse; low light will result in poor flowering and spindly growth.

In outdoor plantings, brachycome likes the same conditions as in the greenhouse. Plant them in full sun, but provide some afternoon shade if temperatures are extremely high.

If not, water stress can pose a problem. Avoid full shade plantings as this will produce plants with lots of vegetative growth but poor flowering.

Irrigation

As mentioned earlier, brachycome likes a well-drained, porous growing medium. Keep the growing medium evenly moist. If plants wilt badly, flowers may drop.

Nutrition

Brachycome is a very light feeder. The following feed rates and schedule will produce maximum flowering and moderate vegetative growth: Apply a constant liquid feed of 75 ppm N, 50 ppm P_2O_5, and 75 ppm K_2O.

Both overwatering and overfeeding will cause abundant vegetative growth, but little flowering.

Scheduling

4-inch: Depending on season and location, crop time to produce a 4-inch plant from a 72-cell pack liner will vary from eight to 12 weeks. Pinch plants either at potting time or one week later to encourage side branching. Plants will fill out the pots rapidly. Cultural information says that B-Nine is effective in controlling plant growth and will also cause the foliage to be darker green.

Space 4-inch pots at a finished spacing of 6-inch by 6-inch or 7-inch by 7-inch so plants will shape up nicely, but not grow together.

Hanging baskets: Plan on a 10 to 12 week crop time for an 8-inch hanging basket with three starter plants, or a 10-inch basket with four starter plants. Obviously, add more production time to the schedule if fewer plants are used.

Insects

Brachycome is fortunate not to be bothered by too many insect pests. However, aphids can appear on occasion. Cultural information states that pest control can be handled satisfactorily on a see-and-treat basis.

Orthene can be used to control aphids.

Brachycome

Diseases
Brachycome has few, if any, disease problems. Under cloudy weather, Botrytis can appear in dense canopies of foliage (which might be present in large hanging baskets).

Control Botrytis with Chipco, Benlate, or Termil.

99

Teresa Aimone, former editor of GrowerTalks *magazine, is with Sluis and Groot, Fort Wayne, Indiana.*

Cultural information for this article was supplied by Mikkelsens, Inc., Ashtabula, Ohio. Plant material is available from Mikkelsens, Inc.

Caladium

Culture notes

April 1985

by Teresa Aimone

Caladium
Family: Araceae
Genus, species: *Caladium* x *hortulanum*

Caladiums are one of the few pot plants that are grown strictly for the foliage. The beautiful patterns of red, pink, white, and green on the large, heart-shaped or lance-shaped leaves more than compensate for the lack of spectacular, showy flowers.

Provided enough heat is available to them, caladiums are a good spring crop to complement bedding sales; they can also help turn profits during the summer when caladiums can have all the free heat they want and empty greenhouse space may be available. The cut leaves also add spice to floral arrangements.

Handling tubers

Caladiums grow from specialized stems called tubers. When tubers arrive, unpack them immediately and lay them either on racks or flats (one layer of tubers only) to provide good air circulation. Inspect the tubers carefully! If they are rubbery, cold damage has probably occurred. The tubers will most likely sprout, but, depending on the extent of the damage, crop time can be dramatically increased.

The purchased tubers should have been stored for at least six weeks at 70 F. Vigor decreases if storage is more than 16 weeks. Tubers stored between six and 16 weeks will be salable in anywhere from four to 12 weeks after planting. The cropping time depends on when the tubers are planted; a winter planting will be ready in 12 weeks, and a May planting should be finished in four to six weeks.

When choosing tuber size, it is important to consider what pot size will be used. A No. 2 tuber (1- to 1½-inch diameter) will provide the proper shoot-to-pot ratio for a cell in a 6-pack; a No. 1 tuber (1½- to 2½-inch) should be used in 3-inch, 3½-inch, and 4-inch pots; Jumbo tubers (2½- to 3½-inch) are fine for 5-inch, and Mammoth tubers (3½-inch to 4½-inch) are recommended for pot sizes larger than 5-inch.

Caladium tubers have many nodes or "eyes" on the surface, just like a potato. Generally, one eye will be larger and more dominant than others on the same tuber. To provide a fuller, more branched plant, it's a good idea to ream out or remove this eye before planting and allow the lateral eyes to grow. The

Caladium

dominance of this larger eye can also be reduced by making a ¼-inch X through this eye. Some growers dust the tubers with sulfur or Captan after removing the eye; others allow the area to heal for 24 hours before planting. If large leaves are desired, don't remove the eye. Removal results in more, but smaller, leaves.

Planting

If the above tuber/container recommendations are followed, plant one tuber per pot. Several smaller tubers can also be planted per 5-inch or larger pot. In either case, plant the tuber approximately 1 inch beneath the soil surface. Use a light, well-drained mix such as soil-less mix or a 1:1:1, soil, peat, and sand. Keep the soil moderately moist; do not allow it to become excessively wet or dry.

Forcing

Caladiums should never be stored, handled, or shipped below 70 F. And temperature becomes especially critical during the sprouting or forcing period. Caladiums can be forced either in their final containers or in beds. Either way, provide caladiums with all the heat you possibly can; bottom heat is very helpful. If it's not available, the post should definitely be tented with black poly until the sprouted leaves begin to unfurl. A combination of bottom heat and tenting will ensure growers that the proper temperature requirements are being met. Plant bed-grown caladiums in their final containers two to three weeks after the start of forcing.

Forcing temperatures for caladiums are 80 to 85 F. Do not allow night temperatures to go below 60 F. Maintain 90 percent humidity under the tent. Periodic misting will help maintain humidity. As soon as the leaves start to unfurl on the sprouted plants, remove them from the tented area and grow on at 70 F.

GrowerTalks on Crop Culture

Caladium

Fertilizing

Caladiums respond well to fertilization, though they do not require large amounts of feed. A top dressing of 14-14-14 Osmocote (one teaspoon per 6-inch pot) supplemented with a liquid feeding of 300 ppm nitrogen every two to three weeks suffices. For the liquid feed, use 20-5-30 or 20-20-20 at the rate of 20 ounces per 100 gallons. Substitute calcium nitrate and potassium nitrate during the cold winter months. If Osmocote is not used, feed weekly with 400 ppm N, P, and K.

Light intensity

Caladiums can be grown under footcandle intensities ranging from 2,500 to 5,000. Lower levels can cause undesirable petiole stretching and color fading. Extremely high light levels can also cause fading. There are some varieties that can tolerate either full sun or partial shade.

Diseases

The biggest problem with caladiums is bulb rot caused by *Fusarium spp.* Dip tubers prior to planting in Dithan M-45 or Benlate for 30 minutes. Rates for both are one pound per 100 gallons.

Varieties

Many varieties of caladiums are available, so the decision depends to a great deal on your color combination preference. Advances in tissue-cultured caladiums have helped reduce the incidence of an inherent caladium disease, dasheen mosaic virus.

The following is a list of good caladium varieties:

Aaron: creamy white center with green border.
Candidum: white with green veins.
Carolyn Whorton: multi-colored leaves of red, pink, and green.
Crimson Wave: crimson with dark green border.
Dr. Grover: transparent pink with dark green border and bright red veins.
Fannie Munson: bright crimson with scarlet ribs and green border.
Fire Chief: crimson heart, green border, and green splotches.
Freida Hemple: bright red center with green border.
Itacapus: dark red with bright red spots.
June Bride: white with thin green borders; almost translucent.
Lord Derby: transparent rose, dark ribs, and green border.
Mrs. Arno Nehrling: white with green splotches, red veins, and green border.
Mrs. F.M. Joyner: white with pink hue and scarlet midribs.
Pink Beauty: pink leaves with green border.
Pink Cloud: crumpled, mottled pink leaves.
Poecile Anglais: deep crimson with metallic green border.
Postman Joyner: deep, dark red with green border.
White Christmas: solid white between midribs; green border.

"

Teresa Aimone, former editor of GrowerTalks *magazine, is with Sluis and Groot, Fort Wayne, Indiana.*

Calceolaria

Culture Notes

August 1986

by Teresa Aimone

Calceolaria (Pocketbook Plant)
Family: Scrophulariaceae
Genus, species: *Calceolaria hybrida*

Striking color combinations and unique flower form are characteristic of this winter-grown pot plant. Traditionally grown with its companion, cineraria, calceolarias enjoyed a surge of popularity several years ago.

Calceolarias are a 28-week crop when grown from seed. Sow in September for March and April flowering. Normally this crop can't be pulled bench-run; flowering is spread out over a 2 to 4 week period. Expect to take 50% of the crop off the bench at first flowering.

Propagation

Calceolarias are normally grown from seed. Seed are extremely small— 500,000 to 1,000,000 seed per ounce. Use a light medium for propagation (something with a great deal of aeration such as 1:1:1, peat, perlite, and vermiculite) and do not cover the seed. Do not give seed full sun. Provide 70 F bottom heat for germination. Germination will occur in 14 to 16 days.

Growing-on

Transplanting: Shift germinated seedlings into 2½- or 3-inch pots or cells and grow pot tight. When leaves touch adjacent ones, plants are ready to shift into 5- or 6-inch pots. Due to their fine, fibrous root systems, calceolarias seem to do better in these smaller pot sizes. Final plant height is also not in proportion to larger container sizes. Watch the watering carefully when shifting to final containers; once again, since the root system is so fine, it's very easy to lose newly transplanted plants due to overwatering.

After plants have been transplanted, place a pot to pot spacing. Space at 10 inches by 10 feet for final spacing.

Temperature: Initial growth should be at 60 F in the early stages. When plants are shifted to final containers, lower temperatures to 50 F. When buds are approximately $1/16$-inch across, raise temperatures to 55 F. Flower buds will not form when plants are grown at 60 F.

Fertilization: Moderate to light feedings are adequate. A common problem with calceolarias is chlorosis at leaf tips. This is caused by: (1) poor root growth in young plants because of poor drainage; (2) very frequent waterings; or (3) too

Calceolaria

Anytime
Red and Yellow
calceolaria

much fertilizer. Use 200 ppm 20-20-20 at every watering. If you use the full rate of Osmocote in your medium, then feed at 200 ppm every third watering, or water with 100 ppm every other watering. During cloudy weather (and if plants won't dry out), provide a calcium nitrate form of fertilizer.

Lighting: Though flower formation is not dependent on light, flowers will develop faster when given long days. When plants have four to five leaves present, provide long days by increasing daylengths by four hours per day. Give plants 10 footcandles of light approximately 24 inches above plant tops. Light treatment should last for three months.

Growth regulators: Cycocel (chlormequat) will provide height control, but is best used when calceolarias are grown on a fast-crop schedule (see accompanying tables). There are several schools of thought when using growth regulators on calceolarias.

The first school is to spray when flower buds are $1/16$-inch in diameter, apply two sprays (two weeks apart) of 400 ppm Cycocel. Each spray should be at 400 ppm. Expect a 20 percent height reduction using this technique.

Or, apply a single spray of 800 ppm at the same initial time (when plants have a $1/16$-inch flower bud diameter). It will produce a similar plant height, but may cause some marginal burn (foliar phytotoxicity) on the leaves.

The third method is to apply a single 3,000 ppm Cycocel drench at the $1/16$-inch diameter bud stage. The drench will reduce height up to 50 percent; flowering won't occur any sooner than the other two methods. Deeper green foliage will result when using these growth regulator methods.

Higher rates of Cycocel applied as either a drench or a spray will produce foliar burn. Try a small group and see how they respond before applying the chemical to the whole crop. The drench may result in some "midget" plants.

Calceolaria

Production stage	Timing	Temperature (F)	Lighting
Sow seed	Early September	65 to 70	Natural daylength
Transplant to flats	Late September	60 to 65	Natural daylength
Transplant to 2¼" or 3"	Early November	55	Natural daylength
Transplant to final containers	Early December	55	Natural daylength
Begin cool treatment	Early December	45 to 50	Natural daylength
End cool treatment	Mid-January	55	Natural daylength
Flowering	Late March		

Table 1. Traditional schedule for calceoarias. Cycocel is not necessarily needed with this schedule since plants are grown slower. Total crop time is 28 weeks.

Production stage	Timing	Temperature (F)	Lighting
Sow seed	Late September	60	18 hours
Transplant to flats	Mid-October	60 to 70	18 hours
Transplant to 2¼" or 3" (begin short days)	Mid-November	60 to 65	8 hours
Begin long days	Late December	50 to 55	18 hours
Flowering	Mid- to late January		

Table 2. Fast crop schedule for calceoarias. Plants should have 4 to 5 leaves before short days are started. Since plants are being grown at warmer temperatures, it's a good idea to use Cycocel. Total crop time is 17 to 19 weeks.

Insects and diseases

Insects: Aphids, thrips, and whiteflies all attack calceolarias.

Diseases: Stem rot (caused by Pythium or Rhizoctonia) may attack calceolarias. It's aggravated when seedlings are planted too deeply, given too much moisture, or provided with too much drainage.

"

Teresa Aimone, former editor of GrowerTalks *magazine, is with Sluis and Groot, Fort Wayne, Indiana.*

Culture notes

December 1987

by Ron Adams

Campanula
Family: Campanulaceae
Genus, species: *Campanula isophylla*

Campanula isophylla belongs to the family Campanulaceae, a genus of nearly 200 species of mostly perennial herbs with a few annuals and biennials.

A native of the Italian Alps, *Campanula isophylla* requires 16-hour days to initiate flower buds. Historically this crop was started from cuttings until two F_1 hybrid varieties were introduced: Stella White and Stella Blue. Plants grown from the Stella varieties are compact with uniform plant habit, with green heart-shaped foliage, large and showy star-shaped flowers and more vigorous plant growth. By starting from seed, production can be programmed for year-round flowering. Crops sown in July will flower in April; crops sown in November will flower in June. For natural season flowering, sow in January for September flowering.

Germination: Sow seed in a peat-lite media and do not cover. Maintain soil temperatures of 60 to 65 F. Supplemental lighting after emergence will develop a stronger seedling that will take off faster, use a minimum of 250 footcandles for 14 hours per day. Relative humidity should be kept at 85 percent to 95 percent. Germination takes up to 21 days.

Forcing schedule

Spring flowering	
Sow	7/1
Transplant	8/1-8/15
Pot into 4-inch or hanging baskets	11/1
Flower	4/1

Summer flowering	
Sow	11/1
Transplant	12/1-12/15
Pot into 4-inch or hanging baskets	3/1
Flower	6/1

Natural season flowering	
Sow	1/1
Transplant	2/1-2/15
Pot into 4-inch or hanging baskets	5/1
Flower	9/1

Transplanting: As soon as seedlings are large enough, in four to six weeks, transplant into a 4- or 6-cell pack. Grow on at a 60 F night soil temperatures. After plants are established, the temperature can be lowered to 50 F night soil temperature, but doing so will add four to six weeks to the overall crop time. Some growers will split the crop between the 50 and 60 F temperature to stagger flowering. Feed with a 15-16-17 peat-lite fertilizer at a constant 200

Campanula

ppm to 250 ppm N rate. The soil mix should be porous to allow good drainage and an air porosity of 25 percent to 30 percent. It is critical not to over water. Keep the soil dry to moist but not saturated.

Potting: Before the seedlings become root-bound, they should be shifted to 4-inch pots or hanging baskets (approximately 12 weeks). Use three plants in an 8-inch basket and four to five plants in a 10- or 12-inch basket. Continue to grow the crop at 60 F night soil temperatures. Again, growing at 50 F night soil temperatures will add four to six weeks to crop time. Pot transplants into a well drained peat-lite media (25 percent to 30 percent soil air porosity) and do not over water. Fertilize with a 15-16-17 peat-lite fertilizer at a rate of 200 ppm to 250 ppm N.

Forcing: *Campanula isophylla* can be forced into bloom since they need long days to flower. Use four hours of supplemental lighting in December and January, three hours of supplemental lighting in November and February and two hours of supplemental lighting in September, October, March and April. This is best accomplished as a night break using mum lights starting four weeks after the seedlings are shifted to 4-inch pots or hanging baskets and until buds are well developed (usually about four weeks).

Growth regulations: Apply B-Nine at a rate of 2,500 ppm two to three weeks after potting transplants in pots or hanging baskets and repeat six weeks later.

Disease and insect problems: *Campanula isophylla* is highly susceptible to Botrytis. Treat with Benlate, Chipco 26019, Daconil or Exotherm Termil.

Aphids, spider mites and fungus gnats are the most troublesome pests to watch for.

99

Ron Adams is technical services manager with Ball Seed Co., West Chicago, Illinois.

Chrysanthemums

Culture notes

September 1990

by Ron Adams

Marguerite daisy
Family: Asteraceae (Compositae)
Genus, species: *Chrysanthemum frutescens*

Marguerite daisy has been a popular cut flower, bedding plant, and in warm climates, a shrub for many years. Marguerite is a white daisy while Boston has a yellow flower. Unlike standard pot, cut and garden mums, marguerites are native to the Canary Islands and require long days to flower; they remain vegetative under short days.

The marguerite daisies are either white or yellow with single florets. Other daisies that are pink, blue, double flowered or dwarf strains typically are of other chrysanthemum species. Outdoors marguerite daisies perform well, although plant habit varies in different areas.

Cut flowers
- **Propagation.** Cuttings for cut flowers are often started in March and grown in 3-, 4- or 5-inch pots depending on when they are to be benched. *Commercial Flower Forcing* recommends spacing of 18 inches per plant in the cut flower bed.
- **Flowering.** *Commercial Flower Forcing* suggests 50 F night temperatures for December flowering. More recent research shows that growing at 62 F nights starts flowering as early as 8 weeks after starting long days. Use mum lights for long days to increase flower production. Pinch out buds during vegetative growth until the plant reaches desired size.

Bedding plants
For bedding plants start cuttings in March. For cell pack production, plants will bloom faster if they are root bound before transplanting. Feed 100 to 200 parts per million N from a complete fertilizer such as 15-16-17 or 20-10-20, one to two times a week.

Problems and postharvest
Insects: leaf rollers, aphids, thrips, leafminers, spider and cyclamen mites.
Diseases: Botrytis, Pythium, Rhizoctonia, crown gall, powdery mildew, aster yellows.

Chrysanthemums

Marguerite daisy

Lighting for flowering

Month	Lighting duration
April and May	2 hours per night
June, July and August	Natural lighting sufficient
September and March	3 hours per night
October through February	4 hours per night

Minimum light requirements

Light bulb size	Bulb spacing	Height above plant
60 watt	4 feet	2 to 6 feet
100 watt	6 feet	4 to 6 feet
150 watt	10 feet	6 feet

Lighting can be as an extended day or a middle of night treatment.

Postharvest: Cut flowers have an excellent shelf life.

You can extend marguerite's postharvest life by using a combination of deionized water with 320 ppm citric acid and 200 ppm quaternary ammonium salt compound.

99

Ron Adams is manager of technical services, Ball Seed Co., West Chicago, Illinois.

Chrysanthemums, garden

Culture notes

May 1989

by Ed Higgins

Garden mums
Family: Compositae
Genus, species: *Chrysanthemum morifolium*

Just as poinsettias color the Christmas season red, garden mums paint late summer and autumn with a palette of colors reflecting the beauty of nature. In addition to their colors and diverse flower forms, garden mums are easy to grow and market in their natural season, the fall.

Varieties upgraded
Recent garden mum introductions vastly upgrade the appearance and performance of the crop. Bright flower colors, new flower forms, improved branching and better uniformity of response are traits found in new garden mum varieties. They are also more flexible in their usage. Most newer cultivars are adapted both to container and field growing and can usually be grown in hanging baskets. For consumers, they are excellent in garden settings, patio planters, window boxes and as gifts. For landscaping, the colors and uniformity of the plants help ensure an attractive display.

When selecting varieties, the knowledge of sales representatives and cutting suppliers can be invaluable. Consider the growth traits that you desire and also whether an early, midseason or late flowering response is needed. Simply by choosing a cultivar, the future quality of the crop is greatly impacted.

Fertilization varies
Garden mums are heavy feeders. The type of fertilizer is not critical, but a reliable, consistent source is. Fertilizer rates vary depending on the type of media and the fertilizer used along with the frequency of application.

Always start a garden mum crop with liquid feed. As soon as cuttings are planted, feed with 300 to 400 parts per million from a balanced fertilizer, such as 20-20-20 or 15-15-15. This initial feed helps young plants get established.

Many fertilizer programs are used successfully for producing garden mums. Liquid, slow release and granular fertilizers all help to produce healthy, quality garden mums. The critical factor is to maintain fertility levels until just before color shows in the flower buds.

Water carefully

Garden mums do not like to dry out. Wilting down early in the season can reduce branching, harden off the plants and contribute to premature budding.

Drip tubes are preferred for container crops. Tube watering helps keep foliage dry and facilitates the use of liquid fertilizers. Overhead irrigation is easy to set up and use, but requires more water, is not efficient for liquid fertilizers and if used late in the day or in the evening, may contribute to foliar diseases, such as Botrytis and bacterial leaf spot.

Desired finished quality and selling price play important roles in determining spacing. Six-inch pots for natural season fall crops may require spacing of 14- to 16-inches on center. Seven- and 8-inch containers may require spacing of 18- to 24-inches on center. Field-grown plants often require spacing of 24- to 30-inches on center.

Garden mums grown too closely together develop a stove pipe or upright appearance and have less value and use than mounded, cushion-shaped plants.

Other cultural factors

- **Medias must be well drained.** Poorly drained medias, whether soil-based or soil-less, may lead to restricted root systems, poor plant growth and often to root diseases, such as Pythium.
- **Containers for fall garden mum production are diverse.** A rule of thumb is the larger the container, the larger the finished plant size. Common containers for all garden mums are 6-, 7- and 8-inch azalea pots, 8- by 5-inch mum pans, and 1- and 1½-gallon nursery cans.
- **Pinching helps produce a full, well branched plant.** Cuttings planted in May and June for September sales are usually pinched twice. The first pinch is usually done two weeks after planting, the second approximately three to four weeks after the first. Pinching should be completed by July 15 in the North and by late July in the South.
- **Garden mums are so reproductive**, the occurrence of buds prematurely can be expected. To minimize budding, always pinch hard (remove one-half to 1 inch of new growth). Also, keep the plants well watered and well fertilized to maintain active growth.

Chrysanthemums

• **Pests are few.** Diseases can usually be prevented with proper growing techniques. Use a well-drained, disease-free mix and don't water overhead late in the day. Proper spacing for air movement is also important.

Fast crops of garden mums are becoming more popular. One cutting is usually planted in a 6-inch pot the last week of July. The cutting should never be pinched. Just "push" the plant with feed and water, and they will branch out on their own because of nature's crown bud process at this time of the year. Well-grown, quality plants can finish with a 12- to 15-inch head size. Fast crops flower approximately four to seven days later than natural crops.

Labeling garden mums is important. Gardening articles in magazines and newspapers mention variety names or series names such as the Prophets. Help to satisfy the customer and use labels for each of your garden mums.

"

Ed Higgins is new varieties manager for Yoder Bros. Inc., Barberton, Ohio.

Growing ideas

September 1985

by P. Allen Hammer

"Have you as a grower made every effort to make your garden chrysanthemums vegetative?"

This past summer I received several calls from growers in June and early July concerned with premature flower buds in their hardy chrysanthemum crop. This seems to have been a problem for growers, particularly in the Midwest this past growing season. I am sure other growers from around the country experience similar problems from time to time, so we will review the possible causes and offer solutions to avoid premature flower initiation.

Short days and long nights

Hardy garden chrysanthemums are short-day plants, which means they require short days, or more correctly, long nights to flower. We generally say the critical day length is $14\frac{1}{2}$ hours for flower initiation and $13\frac{1}{2}$ hours for development. This simply means that the plants will initiate (set) flower buds when they receive $9\frac{1}{2}$ hours or More of continuous darkness during the night period. After initiation, the plants require $10\frac{1}{2}$ hours or more of continuous darkness during the night period for normal flower development. Remember that "crown buds" are an example of initiation without proper flower development because of improper photoperiod control, or a daylength longer than at critical for flower development.

However, hardy garden chrysanthemums have been selected to flower *early* in the fall, particularly in northern areas of the United States to avoid early winter freezes. This selection has been away from those plants that respond to photoperiod as nicely as greenhouse chrysanthemums. In fact, it is very difficult, to almost impossible, to keep some of the garden chrysanthemums vegetative with just photoperiod control. Cultural procedures are used to supplement the long day response.

Cultural tips

Garden chrysanthemums are generally potted in May for fall flowering. The day length from sunrise to sunset is 14 hours 32 minutes on May 12 at 42 degrees north latitude (Boston), so artificial lights to extend the daylength are not generally recommended. Pinching and culture procedures that promote rapid growth are used with the natural long days to maintain vegetative growth. The cuttings should be pinched approximately 10 days after potting. Additional pinches every 10 to 14 days should be made until early or mid-July. This, along with high nitrogen (300 ppm) and potassium fertilization, will help to avoid premature flower buds.

It is also important to avoid any water stress to the plants during the vegetative growth stage. Wilting from water stress will promote premature flower initiation in hardy chrysanthemums. In fact, withholding water has long been a practice to promote flowering in the other crops. However, I personally do not recommend water stress as a means of promoting flowering in any crop.

We do need to remember that many of the modern potting mixes are extremely well-drained and probably have less water-holding capacity than the heavier soil mixes of the past. Even though you may think that the mix is well-watered, the plant may under some water stress. Syringing on those very hot afternoons will also help by lowering leaf temperatures and reducing the evapotranspiration loss.

Budded cuttings

Several growers have also complained about receiving cuttings that were already budded. Certainly the propagator should provide good vegetative cuttings to the grower. As I mentioned before some of the cultivars are extremely difficult to maintain vegetatively, particularly when the stock plants get old or the cuttings are taken from old shoots. If cuttings do have buds when they are received, a hard pinch at potting is recommended. These plants must also receive excellent growing conditions to stimulate new vegetative growth. In any cases, it is very difficult to get vegetative growth from cuttings after have initiated flower buds. It would be useful to make note of the budded cultivars and discuss it with your supplier.

What does all this mean in simple terms? As a grower, buy good quality hardy chrysanthemums adapted for your growing area. Inspect the cuttings when they are received so you know the quality at that stage. Pot the cuttings in a nutrient-enriched media and begin a good fertilization and watering program from the day of potting. Keep the plants pinched to maintain vigorous vegetative growth. Avoid as much stress as possible. Syringing the plants on hot days will reduce stress.

In summary we can say—keep them growing with water, fertilizer, and pinching.

"

P. Allen Hammer is professor of floriculture, Purdue University, West Lafayette, Indiana.

Codiaeum, Croton

Culture Notes

January 1985

by Janet Langefeld

Croton
Family: Euphorbiaceae
Genus, Species: *Codiaeum variegatum, pictum*

Colorful crotons come in a spectrum of colors and patterns, and, being extremely versatile, are in demand for use in tropical landscapes and interiorscapes. Originally from the Indo-Malayan Islands, crotons made their United States debut in 1804—compliments of British traders. U.S. breeders, therefore, have had the past 180 years to work with croton varieties, and the results have created more colors and more patterns of colors on leaves—yellows, reds, and oranges with and without spots, speckles, fishbone patterns, and other marginal coloring. Crotons can have narrow linear-shaped leaves, broad leaves, very small leaves, corkscrew-shaped leaves, or oak-shaped leaves. There are other varieties that produce interrupted types of leaves, where the leaves are broken into two parts and connected by the mid-veins.

Crotons are actually tropical shrubs, and although they do produce inconspicuous flowers, it's the brilliantly colored leaves that make the foliage plants unique—something a little different from typical green foliage plants. In mass displays, either indoors or out, crotons make a fantastic show.

To propagate or not to propagate?

Growers can either start their own crotons from stock plants or buy in cuttings from offshore suppliers. The advantage to propagating your own stock plants is that cuttings will be automatically acclimatized to the growing environment. Cuttings purchased from other countries sometimes experience acclimatization problems. Sometimes rooted cuttings purchased from outside sources will lose their roots and have to grow new ones.

There are advantages to buying in offshore cuttings, however. First, growers will have more room for production by not tying up valuable space with stock plants. Some croton varieties really hog up space and produce very few cuttings per square foot. Purchasing cuttings from other countries also saves growers a lot of time and labor, since cuttings come ready-to-go. Thirdly, it's easy to buy in cuttings. Just pick up the phone, place an order, and you have your cuttings within a week.

But, as mentioned, acclimatization of cuttings purchased from other countries can cause trouble. The cuttings might have been grown under full sun, and U.S. growers who buy these cuttings might experience root problems and some defoliation; cuttings grown in full sun will have a hard time acclimating to growing in lower light conditions. Nutritional problems, and disease and insect invasions sometimes occur, too.

Methods of propagation

Crotons can be produced sexually—from seed—but the process of producing seed is too slow and expensive to make sexual propagation a viable option. Asexual propagation is definitely the way to go, and tissue culture and vegetative propagation are the asexual options. We'll eliminate tissue culture propagation as an option for our purposes, however, since very little tissue culture work is being done with crotons. That leaves vegetative propagation as the best answer.

Growers propagating their own stock plants will need to allow adequate bench space for the plants. Small-leafed varieties will produce more cuttings per square foot than larger-leafed varieties. Adequate spacing for stock plants will allow for good air circulation, which will help in disease control, and sunlight will be able to penetrate to the lower foliage if enough space is provided. The more sunlight the lower foliage receives, the more cuttings a plant will produce. Raised benches are recommended for stock plant production.

Tip cuttings, joint cuttings, and air-layered cuttings can be taken from stock plants. Tip cuttings are used most frequently because once the cuttings are stuck and develop roots, they're soon marketable plants. Joint cuttings will produce the same good quality plants as tip cuttings, but turnover time is longer. Air layering is recommended only for crotons that will be marketed in larger containers—at least 5-inch and up—since air-layered cuttings are more expensive to produce than tip cuttings. Air layering is most often used to propagate large-leafed varieties in 14-inch and 17-inch containers (a lot larger than the 5-inch pots initially mentioned!).

It is very important to take only colorful cuttings from stock plants because what you see is what you get: The color of the cutting determines the color of the final plant, so it's necessary to develop the color on the stock plants. A dull-colored cutting will have to grow 24 inches or taller before it has another chance to develop good color.

To develop good color, stock plants should receive a maximum amount of light, but full sun is to be avoided. 2,000 to 3,500 footcandles is a good range; if more footcandles are provided, they won't serve any additional benefit. Plants grown under 50 to 100 footcandles will survive, but the color will be poor. Spacing comes into play here, too, because the more light that penetrates to the lower foliage, the more colorful cuttings produced by the plants.

The propagating environment

Croton cuttings can be stuck and rooted in pots in tent beds or open-air beds with or without bottom heat. Tenting probably works best, since 100 percent humidity can be maintained under the plastic tents. Also cuttings rooted in

Codiaeum

tent beds can root up to 1½ weeks earlier than cuttings propagated in open-air beds. The perfect rooting environment combines the use of tent beds and bottom heat. Your cuttings will love it, and probably develop roots in about two weeks.

Spacing: In propagation beds, proper spacing is again important because disease (such as Rhizoctonia) will spread rapidly with the high humidity and high temperatures (particularly in tent beds).

Media: Any well-aerated medium that drains well can be used to propagate crotons. Media for stock plant and cutting production should have a pH of 5.5.

Misting: Misting is extremely important in rooting croton cuttings. During propagation, crotons—especially large-leafed varieties—should always have a good layer of mist on their leaves. Intermittent misting works well in propagating cuttings. Depending on the time of year, 15 minutes of misting per hour or 15 minutes every half hour will do the job. More misting is required in winter when heaters are running. In many areas of the United States, humidity is naturally high in the summer so misting can be more infrequent. Growers might want to continue misting plants into the evening (7 p.m. or so) in the winter, but in the summer misting can end at about 5 p.m.

Temperature: Crotons grow best when temperatures are maintained between 75 and 80 F.

Lighting: As discussed earlier for stock plant production, 2,000 to 3,500 footcandles of light provide optimum lighting conditions for propagating cuttings, too.

Watering: Media should be kept moist, but not saturated with water. Late evening watering can spot leaves, so it's best to water in the morning.

Fertilization: Crotons experience two stages of growth: one from the time the cuttings are rooted until the plants begin to produce breaks, and the second to harden off plants. In the second phase, leaf color will mature and leaves will thicken. In the initial growth stage, a 20-20-20 fertilizer should be used; in some cases, Osmocote or another slow-release fertilizer can be added as a topdress to the roots. In the second growth stage, the nitrogen level of the fertilizer should be reduced to harden off the tissue for good thickness and to properly develop the bright colors. An 18-20-20 reduced nitrogen fertilizer works well in this stage. Crotons like all the fertilizer they can get in the initial growth stage, but fertilizer applications must be reduced in amount applied and frequency of application in the hardening off stage.

Preventative programs: After cuttings are stuck, any good preventative fungicide should be applied as a drench; apply a fungicide once or twice a month thereafter to protect against disease. Truban or Benlate are good fungicides to use. In conjunction with the fungicide, a miticide should be applied every one to two weeks after cuttings are rooted to prevent the invasion of insects. Kelthane with Pentac works well.

Growth regulators: Growth regulators are not necessary but will help control height if plants cannot be immediately shipped upon maturity.

Diseases

Crown gall, a bacterial disease that causes the tissue of a cutting to swell, can be controlled by using sterilized cutting tools. Growing or purchasing disease-free stock is the best prevention against the infrequent problem of crown gall.

Anthracnose is another infrequent problem of crotons and can easily be controlled with a good fungicide. The disease is caused by a fungus and appears as water-soaked lesions on the croton leaves. Too much moisture can bring on the disease; early morning watering can prevent anthracnose from occurring in the first place. Rhizoctonia is also brought on by high humidity, but unlike crown gall and anthracnose, Rhizoctonia comes to visit crotons much too frequently. If Rhizoctonia is not spotted and controlled immediately, it can spread throughout an entire croton crop in two or three days. Tissue will yellow and eventually turn brown and collapse unless treatment is rendered. If a grower using tent propagation notices an invasion of Rhizoctonia, he should lift the sides of the tent to circulate the air and apply a good fungicide as a drench.

Insects

Mealybugs can cause a stunting of terminal growth in crotons. Their appearance can first be detected as white cottony masses in leaf axils and on the leaf undersides. Use a good contact insecticide to get rid of mealybugs, then follow up in a few days with a systemic insecticide. Always use a spreader sticker to apply these insecticides, so the chemical will penetrate the waxy coat protecting the mealybugs.

Scale causes much the same type of damage as mealybugs. Plants will experience a vigorous decline, first yellowing and later defoliating. As with mealybugs, use a contact insecticide followed up by a systemic.

Thrips occasionally bother crotons and will cause distorted terminal growth. It's hard to detect their presence, but small silvery areas on the leaves will signify the presence of thrips. Any insecticide labeled for greenhouse thrips should get rid of them.

Mites are the number 1 problem on crotons. Tissue will yellow, eventually turning brown and then dying. The colorful leaves of crotons often hide the disease in its initial stages, however, so the best control is to regularly use an miticide to prevent its occurrence.

Again, with all of these applications, spreader sticker will work best.

Color development problems

If the color of croton leaves is developing poorly, the plants may not be receiving enough light. Also, too much fertilizer, especially nitrogen fertilizer, will cause crotons to grow vigorously, but the growth will be green. Too high temperatures can also cause poor color, but temperatures above 95 F can be compensated for by increasing the light supply and reducing the amount of fertilizer.

Variety tips

Good linear-leafed croton varieties to grow include Gold Dust. Oakleaf-shaped varieties include Elaine and Bravo, and excellent broadleaf varieties are Norma and Petra. Narrow-leafed varieties include Goldstar, Goldfinger, and Superstar.

Smaller to medium-leafed varieties, such as Goldstar, Goldfinger, Goldsun, Iceton Yellow and the old-fashioned reliable Aucubifolium, are good to use for 2-, 3-, 3½- and 4-inch pots. Norma and Bravo have larger leaves but can be adapted for smaller pot sizes.

The trick to adapting these larger-leafed varieties to smaller pots starts with the stock plants. Put the stock plants in a smaller container than normally called for—for instance, put stock plants in 2-gallon containers instead of 3-gallons. Reduce the amount of nitrogen applied to the stock plants. An increase in light will compensate for the drop in fertilizer and compact growth to reduce the size of the cuttings; color will be better, too. Watering less often will also help reduce the leaf size of the normally large-leafed varieties. By following this method, large-leafed varieties can be adapted to 3- or 4-inch pots.

For pots 5-inch and larger, try Norma, Petra, Bravo, or the Iceton varieties. Just about any variety of croton can be grown in larger pots, depending on when you want to finish the crop since production time varies with size of croton leaves.

In the North, crotons must be grown indoors, of course, and are usually propagated as 4- to 12-inch potted plants. Outdoors in southern Florida, crotons will mature anywhere between 10 and 15 feet tall. At this size, they make extremely attractive landscapes, but smaller sized crotons can add a splash of much wanted color to any interiorscape.

99

Janet Langefeld is a former GrowerTalks staff writer.

Cotula

Culture notes

January 1989

by Teresa Aimone

Cotula
Family: Compositae
Genus, species: *Cotula turbinata*

This unusual plant has pompon-shaped, golden-orange flowers. Flower shape is similar to *Cotula barbata* or pincushion plant.

Cotula can be sold as a bedding plant in packs or pots or used as a short summer cut flower. This variety is for outdoor cut flower production only. It can also be used as a dried flower; wiring the stem is recommended.

Germinate seed at 59 to 68 F. Grow-on at 50 to 65 F for six to eight weeks before transplanting outdoors for cut flowers. Cotula plants do best in full sun.

Cotula can be direct-sown into the field in very late spring. Final height varies from 16 to 18 inches with direct sowing in rows to 10 to 20 inches if plants have been transplanted. Crop time is three months.

Cotula can overwinter in mild climates, and it is cold-hardy to 32 F.

99

Teresa Aimone, former editor of GrowerTalks magazine, is with Sluis and Groot, Fort Wayne, Indiana.

Culture Notes

February 1985

by Janet Langefeld

Crossandra (Firecracker flower)
Family: Acanthaceae
Genus, species: *Crossandra infundibuliformis*

Beautiful crossandra plants display glossy, gardenia-like foliage with overlapping clear, salmon-orange florets. Crossandras are considered a minor crop that is usually grown to complement spring/summer fill-in programs. Mother's Day and Memorial Day are good holidays to sell crossandras.

Originally from India, crossandras have been improved through breeding work to be more uniformly floriferous. The most popular variety, Mona Wallhed, is a result of Danish breeding techniques. Crossandras were introduced to the United States in the early 1900s.

Propagation

Most crossandra growers buy in 2½-inch liners to start crops. Plants can be grown from seed, but the procedure is tedious and germination can be sporadic. For best germination, temperatures must be alternated nights (70 F) and days (85 F) for three to four weeks. Seed should be covered with ⅛ inch of medium to prevent drying out. Germination will start 10 to 14 days after sowing and will continue in a slow and erratic fashion for about one month. Transplanting must be done irregularly, therefore, as plants become large enough to handle. When transplanting, be careful not to disturb ungerminated seed.

Crossandras started from seed will take seven to eight months to flower—a very long crop time. The slow and erratic germination plus the long crop time are reasons why growers start plants vegetatively or buy in 2½-inch liners.

Growers propagating their own crossandras from cuttings should take soft tip cuttings from rapidly growing plants. Direct stick the cuttings in 2½-inch pots and root in a warm (70 F nights, 85 F days) air temperature with a soil temperature of 75 F. Use a rooting compound, if desired. A sterilized medium with a good water-holding capacity is needed to grow crossandras. (The sterilized medium will help prevent the onset of Rhizoctonia). Light levels to root cuttings should be between 1,500 to 1,800 footcandles. If possible, root the crossandras under propagation tents or take some other means to keep humidity high. Use a soil drench of Banrot plus one application only of Terraclor (more than one application will cause burning) to prevent fungal infection. Cuttings can be pinched after six weeks to encourage branching, or

Crossandra

plants can be pinched later after transplanting. In the early stages of growth, crossandras must grow as vegetatively as possible—similarly to growing philodendrons or other foliage plants—to prevent early flowering. Feed cuttings heavily to encourage rapid vegetative growth.

Cuttings will take three to four weeks to root and will be ready for transplanting to larger pots after nine weeks. After transplanting, crossandras will finish in 12 to 14 weeks.

Production from 2½-inch

Potting and media: Unpack and pot liners as soon as possible upon arrival. Pot one plant to a 4½- or 5-inch pot or three plants to a 6-inch pot. Potting level should be the same as it was in the 2½-inch liners—not too shallow and not too deep. If plants are potted too deep Rhizoctonia can set in.

Use a well-drained, sterilized medium. Soilless mixes are recommended. After potting, drench with Dexon (Lesan)/Benlate, Banrot, or a similar fungicide. Repeat fungicide treatment every three to four weeks.

Spacing: Place crossandras pot-to-pot for first six weeks, then move to final spacing of 6½-inch by 7-inch for 4½-inch pots and 9-inch by 9-inch for 6-inch pots.

Temperature and lighting: Maintain 70 F night temperature, 78 to 82 F day temperature for the first six weeks. After that, night temperature can be lowered to 65 F. It's imperative that crossandras be grown warm, an at no time should temperature fall be low 45 F. If temperature did drop below 45 F, necrosis will occur: Leaves will turn black, and, literally, the crop will be devastated.

Grow crossandras under an optimum 2,600 to 3,000 footcandles Do not grow below 2,000 footcandle or above 3,500 footcandles. They need bright light, but summer crop will require about a 30 percent shade.

Humidity: Maintain humidity at 70 to 80 percent during the first six weeks after transplanting, then maintain at least 30 to 50 percent.

Pinching: If pinching did not begin at the cutting stage, pinch all shoots four weeks after transplanting to encourage branching.

Feeding and watering: Use a mum feeding program when growing crossandras, and maintain a pH of 6.2 to 7.0. Feed with about 200 ppm nitrogen, phosphorus, and potassium at every watering.

Use tube watering if possible, but do not allow crossandras to dry out as this will cause leaves to burn.

Growth retardants: Use Alar or B-Nine for height control at 2,500 ppm two weeks after pinching final pots, or when new shoots are 1½ to 2 inches long. Growth retardant will probably not be needed in summer.

Flowering: All flower buds should be removed as they become visible after transplanting and during the first two weeks of the finishing period (finish time is six to seven weeks).

Pests and insecticides

Whitefly is the major pest of crossandras but can be controlled by using Resmethrin (SBP-1382) as a foliar spray, aerosol, or fog. For a spray, use one pint of 2EC (emulsifiable concentrate) per 100 gallons of water. Follow label directions for aerosols and fogs.

If crossandras are grown outside, worms may be a problem. Control worms with Thuricide at four tablespoons per one gallon of water, and apply as a spray.

99

Janet Langefeld is a former GrowerTalks staff writer.

Cut flowers

Perennial pickings:
5 garden favorites cut from the greenhouse

March 1990

by Richard R. Iversen

Looking for some perennial cuts to expand your crop base? With the right cold and light treatments, you can cut traditional garden favorites— platycodon, campanula, heuchera, lysimachia and echinops—year-round from your greenhouse.

For Americans without a 60-foot herbaceous border, the coveted English flower garden is found in flower arrangements that convey the English garden look. This new floral design style is achieved with cut perennial garden flowers. Dozens exist; many old-fashioned favorites suggest a country freshness.

Dutch growers supply most of these materials or they are only used when in season. Now, however, American growers can adapt new research methods and produce perennial flowers all year long.

Perennial plant forcing isn't entirely new. *Chrysanthemum morifolium*, the florist mum, is an herbaceous perennial from the Orient. This short-day plant can now be brought into flower almost at will. Bristol Fairy *Gypsophila paniculata*, perennial baby's breath, has become a flower shop staple, and liatris is increasingly found in floral arrangements around the world.

Flowering requirements of all plants need to be understood before they can be forced in the greenhouse. Environmental cues such as cold temperature and daylength often trigger flower induction, those invisible stages that lead to flower formation.

Meeting cold requirements for flowering and dormancy

Many garden perennials are long-day plants. Some also require cold periods to break root, rhizome, tuber or crown dormancy and/or induce flowering. When cold isn't needed, it often accelerates flowering and improves quality.

Platycodon grandiflorus (balloon flower), *Campanula persicifolia* (bellflower), Chatterbox *Heuchera* x *brizoides* (coral bells), *Lysimachia clethroides* (gooseneck loosestrife) and Taplow Blue *Echinops* (globe thistle) all have specific cold requirements. Cold is necessary to break platycodon's crown dormancy. Campanula and heuchera require cold for flower induction. Crown

Cut Flowers

and rhizome dormancy of both lysimachia and echinops are broken by cold; flowering is induced by long days.

Traditionally, chilly nights in a cold frame satisfy cold requirements, but refrigerators or walk-in refrigerated rooms provide more precise timing for floriculture crops. Prior to refrigeration, grow plants under short daylengths for at least one month and drench with fungicide. During refrigeration monitor plants for Botrytis infection and water when necessary.

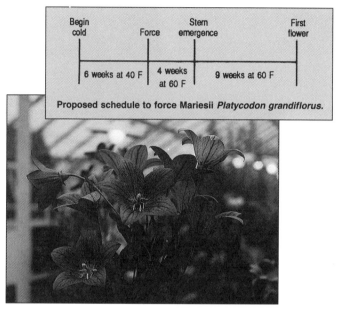

Platycodon (balloon flower)

Lysimachia (gooseneck loosestrife)

Cut Flowers

Chatterbox heuchera

Proposed schedule to force Chatterbox *Heuchera* x *brizoides*.

Begin cold — 12 weeks at 40 F — Force — 9 weeks at 60 F — First flower

Force long-day flowers with lights

Out-of-season flower induction in long-day perennials requires extending natural daylengths in fall, winter and early spring. Use a series of 60-watt incandescent lamps spaced 24 inches apart and 24 inches above the plants to provide about 20 footcandles of light. For 16-hour daylengths, light lamps from sunset to midnight or use interrupted lighting from 10 p.m. to 2 a.m.

Fill greenhouse beds with a root media consisting of $1/3$-bushel sterilized silty clay loam, $1/3$-bushel peat moss and $1/3$-bushel perlite. Amend with 3 ounces of 20 percent superphosphate, 3 ounces of dolomitic limestone and 1 ounce of 10-10-10 fertilizer per bushel. Supplement water with a 20-10-20 peat-lite fertilizer at 150 parts per million nitrogen.

Field grown nursery plants, garden transplants, divisions or mature seedlings are all used for forcing. Highest quality blossoms are produced from large, high quality rootstocks. Like most other crops, growers must stake and monitor for insects and diseases.

Crop by crop specifics

Long, white, horseradish-like taproots of ***Platycodon grandiflorus*** shipped from wholesale suppliers in early autumn require a minimum 6 weeks cold storage before planting in greenhouse beds. Longer cold periods accelerate initial stem growth. Plant taproots with their crowns at soil level. Harvest stems after the first inflated balloons burst into lavender-blue stars.

Campanula persicifolia and **Chatterbox *Heuchera* x *brizoides***, two evergreen species, need light during cold storage for foliage to remain alive. Twenty to 50 footcandles from incandescent lamps for 8 hours per day is adequate. This lighting doesn't affect flower induction. Both need 12 weeks cold storage. Force at any daylength. Sprays of coral bells can replace baby's breath in a bouquet.

Dig *Lysimachia clethroides* rhizomes from garden stock beds or order them from perennial wholesalers. They require 12 weeks of cold storage. Pot, drench with a fungicide, refrigerate, then plant in greenhouse beds. Long daylengths are necessary for growth and flowering. The slender, nodding floral spike, which bows like the neck of a goose, is a unique cut flower.

Taplow Blue *Echinops* is treated similarly to lysimachia, but needs only 6 weeks of cold storage. Twenty-four-hour daylengths produce repeated harvests of these globular, steelish-blue, hedgehog-like flowers. Stems last 2 weeks in the house. When dried for 2 weeks at 50 F they provide lasting winter decoration.

Perennials, once again popular in the garden, exhibit great potential as new cut flower crops. Don't miss your chance to jump on the perennial plant bandwagon and start producing, marketing and retailing cut perennials.

"

Richard Iversen is director of horticulture at the Staten Island Botanical Garden and a former assistant professor of horticulture, State University of New York, Farmingdale. Richard did his doctoral research on perennial plant forcing at Cornell University, Ithaca, New York.

Cyclamen

Culture notes

October 1988

by Teresa Aimone

Miniature cyclamen
Family: Primulaceae
Genus, species: *Cyclamen persicum*

Miniature cyclamen offer growers and consumers a familiar favorite blooming plant in a new product form. These diminutive relatives of standard-sized cyclamen have cultural requirements that are quite similar to their larger cousins and they offer a few added extras as well.

Miniature cyclamen have a shorter crop time (six to seven months), some of the colors are scented (increasing consumer appeal) and more plants can be grown per square foot of bench space (miniature cyclamen are intended for growth in 3-inch to 4-inch pots). Growers who have produced miniature cyclamen are also finding a market for them as the blooming plant in European dish gardens.

Cyclamen

Little Dresden cyclamen

Propagation

Miniature cyclamen are propagated by seed which is sold by count, not by weight. For best results, sow seed in moist, neutral peat moss. Cover seed. Seed can be sown in rows or open flats, or in larger (around 100-count) plug cells. Maintain uniform moisture levels throughout germination. Germination will take approximately three to four weeks.

Optimum germination will be achieved when seed are provided with a constant 63 to 65 F temperature. After germination, young seedlings should be protected from direct sunlight. They should also be misted regularly to provide high relative humidity. Maintain a constant temperature of 65 to 68 F until transplanting.

Growing-on

Transplanting: Transplant into final containers approximately three to four weeks after germination. Seed that has been sown into flats may need to be transplanted a bit sooner if crowding occurs or if you observe seedlings are growing unevenly in the flat.

Young seedlings will have formed small corms at the time of transplanting; the corms should be placed level with the soil surface. Deeper plantings could result in crown rot. The corm will settle naturally into the soil when watered.

Medium: Use a well-drained, sterile medium consisting mainly of peat. Amendments such as sand, vermiculite or perlite may be added to the soil in the quantity of 10 to 15 percent.

Fertilization: After transplanting, begin regular feedings with 150 parts per million of 20-20-20. This is a slightly a lower concentration of fertilizer than used for standard-sized cyclamen. All adaptions of watering, feeding and temperature are meant to keep plants dwarf and compact, so that the miniature plant size is kept in proportion to the container.

Watering: Miniature cyclamen should be grown somewhat drier than standard-sized cyclamen.

Temperature: Maintain a night temperature of 65 to 68 F for five to six weeks after transplanting. After this, night temperatures can be lowered to 62 F. The day temperatures can be slightly higher (5 F). Higher temperatures slow down plant growth.

Toward the end of the culture, night temperatures can be lowered to 55 F. Good ventilation is essential at these lower temperatures. The tighter foliage canopy on miniature cyclamen encourage disease.

Light: During spring and summer, miniature cyclamen can be grown at a spacing of 8 inches by 8 inches by 9 inches.

Flowering: When potted in late May through early June, miniature cyclamen will flower from late October on. Common flowering times for miniature cyclamen are Thanksgiving, Christmas, Valentine's Day and Mother's Day.

Varieties

Little Dresden: These varieties are true miniatures xhibiting a perfect balance between symmetrical plant habit and small container size. Some of the colors are scented. There are nine separate colors as well as a Mixture available: Brigitte (pure white), Annelie (white with purple eye), Steffi (lilac rose with darker rose eye), Rosemarie (light pink with darker pink eye), Kati (lilac), Betty (deep rose with darker rose eye), Gabi (scarlet), Violetta (violet) and Walter (red).

Mirabelle: There are six separate colors as well as a Mixture available in this series. Colors are Purple, Lilac, Deep Salmon, Light Salmon, Scarlet and White.

"

Teresa Aimone, former editor of GrowerTalks *magazine, is with Sluis and Groot, Fort Wayne, Indiana.*

Culture Notes

June 1986

by Teresa Aimone

Cyclamen
Family: Primulaceae
Genus, species: *Cyclamen persicum*

Cyclamens, those popular Christmastime plants, are now finding their way into growers' greenhouses for longer and longer periods of time. In the past, cyclamens have found their niche at Christmas for several reasons: They're a cool-loving crop; the bloom colors—red, pink, lavender, and white—fit perfectly into holiday scene. And, during the 19th century, cyclamens were the most popular plant for Christmas before they were replaced by poinsettias.

Cyclamen

There are two different ways to produce cyclamens: a fast crop and a long crop. We'll discuss the more traditional long crop method first.

Propagation

Cyclamens are propagated by seed. Sow seed in September and October to get salable plants for next year's Christmas. Sow seed in a flat of neutral (pH 6.8) peat moss; maintain a temperature of 60 F until seed sprout. After germination, lower the temperature to 55 F Always cover the seed. You can cover the seed with ¼ inch of peat moss or simply turn another flat over the top of the germination flat. Remove the flat when seed sprout. Cyclamen seed can also be sown in peat pots.

Growing-on

Planting: After two or three leaves have developed, the young plants are ready for transplanting. Plants may be placed in 2½-inch pots, but they may tend to dry out quickly, so planting them in 3-inch pots, or 3 inches apart in flats or benches is preferred. Apply weekly applications of 20-20-20 at 100 ppm while the plants are still small.

Transplant into larger containers after plants have grown. Some growers gradually shift plants into larger containers as plants enlarge. Cyclamens are relatively low-growing plants, so transplanting them into 7-inch and larger containers might put them out of proportion in relation to the size of the pots.

When planting, always keep the crown of the corm above the soil surface. Crown rot may occur with deeper plantings.

Medium: Soil mixes should consist of mostly non-acidic peat (sphagnum peat has a lower pH than muck peat or Michigan peat), or leaf mold with ¼ to ⅓ soil added. If this mix is used, single superphosphate (0-19-0) should be incorporated into the mix. If you use triple superphosphate (0-46-0), then you need to add one pound of gypsum per cubic yard. This is because sulfur is necessary for plant growth, and triple superphosphate contains no sulfur. Single superphosphate does.

Other mixes commonly used are a 2:2:1 combination of soil, non-acidic peat, manure, or a mix of 1:5 loam and sphagnum peat.

Fertilization: High fertilizer levels should be avoided because they result in heavy vegetative growth and no flower bud development. High levels are to be avoided especially at the beginning and ends of production.

Temperature: Experiments indicate that 59 F is the optimum night temperature for growing cyclamens. Temperatures during the late fall affect the flowering date. While 50 F may have been found to be optimum, 55 F will make them flower sooner, but with no increase in quality. Avoid temperatures of 70 F and higher, if possible.

Light: During the summer, plants should be grown in rather heavily shaded greenhouses (provide 3,000 to 4,000 footcandles of light) and spaced so leaves do not touch. High light will make the plants stretch. In late September, remove the shadecloth.

Watering: Frequent syringing is beneficial for cooling as well as for mite control. Mites like hot and dry conditions. Keep the soil constantly moist, but avoid standing water. Wilting will check plant growth, but standing water, especially in the plant's crown will cause rot.

Flowering: Flower initiation is a combination of several factors:
• Plants flower best when pot-bound. A 5-inch pot-bound plant will flower better than a 6-inch plant that isn't pot-bound.
• Maintain a low level of fertilizer, especially nitrogen. Some recommendations say not to feed after the initial potting. Others say to quit feeding in

October, wait for the buds to set, then resume feeding.
- Maintain a high light intensity—this is especially helpful during flowering.
- Reduce the watering at the onset of buds. This will reduce vegetative growth and promote flowering.
- Maintain night temperatures of 50 to 55 F.
- Apply gibberellic acid (GA) if desired. Flowering can be advanced by spraying plants with a 25 to 50 ppm spray of GA. Apply the spray two to two and a half months before the desired flowering date or when the inflorescence has grown 1 to 1½ inches. Insert the spray nozzle below the leaf canopy to ensure wetting the flower ends. Applications above 50 ppm are detrimental to the plant. Remember, GA only elongates cells, it doesn't make more cells, so higher concentrations will make the portion of the plant that supports the flower weak. Hence the flower will not stand up straight as it should. You may wish to make an initial spray of 25 ppm, and then make further applications based on the effectiveness of the initial spray.

Insects and diseases

Insects: Cyclamen mites are the real problem on cyclamens. These microscopic bugs feed on new plant growth, buds, and unfolding leaflets. They cause malformed leaves. Control cyclamen mites with Kelthane 35 WP, using a concentration of one pound per 100 gallons every 10 to 14 days. Thiodan 25 WP, at a concentration of two pounds per 100 gallons, is also effective. Apply at the same intervals as Kelthane.

Nematodes will also attack cyclamens. Destroy infected plants, and plant seed and corms (the "tuber-like" structures cyclamens form) in sterilized soil. If you are using leaf mold, manure, or soil in your mix, nematodes can be a problem if the soil isn't sterilized.

Diseases: For Botrytis control (*Botrytis cinerea*), spray with Zineb 75 WP weekly at a concentration of one and a half pounds per 100 gallons. Provide good air circulation, lower humidity, and remove dead plant parts.

Cyclamens are also attacked by several types of leaf spots. Spray with Ferbam 76 WP or Zineb 75 WP at one and a half pounds per 100 gallons weekly. Give plants good air circulation, reduce humidity, and remove spotted and dead leaves. For petal spot, use Zineb at the above concentration every three to seven days.

For root rot, (*Thielaviopsis basicola*), destroy infected plants and plant clean seed and sound corms in sterilized soil. Labeled fungicides are also effective in control.

Soft rot (*Erwinia caratovota*) will infect the roots, corms, and lower stems. A distinct odor will be produced in later stages. Destroy infected plants, avoid wounding plants and splashing water, and plant clean corms in sterilized soil.

Stunt (*Ramularia cyclaminicola*) will produce curled leaves on cyclamens. Spray weekly with Ferbam 75 WP at the rate of one and a half pounds per 100 gallons. Plant clean corms or seed in sterilized soil.

Fast-crop cyclamens

There are times when you just don't have that production space to give to the long-term crop. So here's a fast method to use. Remember, this method is less forgiving, requires higher temperatures, and insists on strict attention to the production rules.

Sow seed of early flowering cultivars on April 1 for Christmas sales. Sow seed in nutrient-enriched peat moss or peat pellets in flats or similar containers. Cover the seed with a maximum of ¼ inch of peat; place the flats in the dark at 60 to 65 F. After germination, transfer the flats to a 68 to 70 F greenhouse.

Cyclamen

These are all nighttime temperatures. Apply light shade for a few days after seed have germinated.

No fertilizer is needed for 60 days after sowing. After this, fertilize with 20-20-20 at 100 ppm after the first leaf has developed. Maintain 68 F night temperature after this.

After the plants have developed two to three leaves, transplant them to a 3-inch by 3-inch spacing in flats or directly into the greenhouse bench, or into 3-inch pots. Keep the top of the corm above the top of the growing medium.

Around August 1, place in 5-inch pots, using a 9:1 nutrient-enriched medium (peat to soil mix), or a 1:1: 1 peat moss, sand/perlite, soil.

Follow previously mentioned conditions for light, watering, flowering, insects, and diseases.

"

Teresa Aimone, former editor of GrowerTalks *magazine, is with Sluis and Groot, Fort Wayne, Indiana.*

Culture Notes

December 1985

by Teresa Aimone

Dahlia
Family: Compositae
Genus, species: *Dahlia pinnate*

Tuberous-rooted dahlias are versatile plants for the end consumer; they can function as both a flowering houseplant as well as an outdoor garden plant. In addition, the tubers can be redug and planted the following spring.

The dahlias' flowers come in a wide variety of forms—from cactus-type to decorative to peony and anemone. Vivid flower colors (both solids and bicolors) add to this plant's attraction and help make it a winner for spring sales.

Planting
Plant the tubers upon arrival in their final containers and place the crowns of the tubers just above the soil line of the medium—even with the pot rim. Don't split the tubers apart; splitting the tubers will produce an open wound on the tuber surface, thus providing an open entry for disease organisms.

Planting time depends on cultivar and desired flowering date. For mid- to late April flowering, plant tubers from January 26 to February 9. For early May flowering, plant from February 6 to February 20. For flowering in late May, plant from February 26 to March 12.

Medium: Dahlias should be planted in a well-drained medium that is low in soluble salts. Maintain a soil pH of 6.0 to 7.0. The mix composition should be a 1:1:1 combination; examples would be loamy soil, coarse aggregate, and peat, or peat, perlite, and pea gravel. Recommendations state that the mix should not consist of over one third organic material, nor should it contain bark. After planting and watering the tubers in, the level of the medium should be $1/2$ to 1 inch below the rim of the pot.

Containers: Since dahlia tubers tend to occupy a good portion of the container area, a general rule of thumb is to plant one tuber per 10-inch and smaller container. Standard pots are recommended rather than azalea pots, again, due to the size of the tubers.

Spacing: Grow plants pot-tight for the first five to six weeks; after this, space to a minimum of 12 inches on center.

Disease prevention: Upon planting, drench the tubers with a fungicide such as Banrot, Chipco 26019, Lesan/Benlate, or Truban/Terraclor. See labels for recommended rates. Also, do not use the Truban/Terraclor combination more than once, since it can cause phytotoxicity.

Growing-on

Watering: During the first four to five weeks of growth, keep the medium moist, but not wet. Leach weekly, and avoid wetting the foliage as it develops. Water requirements greatly increase as the plant matures; during the bud development stage, waterings may be very heavy and quite frequent.

Fertilization: After shoot growth becomes visible, begin continuous fertilization with a balanced N, P, and K feed. 20-20-20 can be used since dahlias are grown during normally sunny spring days and ammonium build-up shouldn't be a problem. If desired, a slow release fertilizer such as Osmocote 14-14-14 can be incorporated into the mix. Use at the half-rate of five pounds per cubic yard.

Temperature: Force dahlias at a minimum night temperature of 62 to 65 F. Lower temperatures delay flowering; for every day below 62 F, add one day to the crop time.

Maintain day temperatures of 68 to 72 F. Flowering can be accelerated at higher temperatures, but plant quality can also be reduced if high temperatures are maintained over prolonged periods. Avoid temperatures above 80 F; for every day of production time at this temperature, add one day to the total crop time.

If a grower wishes to delay plant development near market dates, dahlias may be held at 55 F nights. On the other hand, flowering can be accelerated if dahlias are grown at 70 to 75 F for a few days.

Humidity: Provide good ventilation. Dahlias do not like to be grown in areas with high humidity.

Light: Provide fairly high light intensities (2,000 to 5,000 footcandles). These light levels help to control height; lower levels of light, even with the use of growth regulators, will produce stretched plants. Recommendations state that dahlias must be exposed to full sunlight throughout the growing time. Do not apply shade to the crop before May 1.

Dahlias are photoperiodic, like many members of the Compositae family are, so naturally increasing daylengths of 10 to 14 hours will provide the earliest flowering. Long-day treatment, like that used on mums, will delay flowering, but not harm plant quality. Short-day treatment will produce poor quality plants and prevent or delay flowering.

Growth regulators: A-Rest (ancymidol) can be used for height control on dahlias. Use this chemical as a drench when shoots are about 1/4 inch long (no later than two weeks after planting). Response to A-Rest will vary with cultivar. The *Holland Bulb Forcer's Guide*, Third Edition provides specific information in regards to both cultivar response and application procedures.

Pinching: Cultivars vary in the number of shoots they develop, and there are some general rules that will allow growers to determine how they should pinch each plant. Pinch all plants that develop only a single shoot. If plants develop two strong shoots, pinch neither shoot. If plants develop two shoots, one strong and the other weak, pinch only the strong shoot. If plants develop three shoots, do not pinch any of the shoots. Pinching should be done after three to four pairs of leaves have developed. Pinching will also delay flowering five to 10 days and will produce taller plants than non-pinched. If the majority of the crop has reached the pinching stage, pinch those that require it and leave the less developed plants unpinched.

Diseases and Insects

Most diseases originate in the propagation field, and viruses are the main disorder carried by dahlias. With the exception of viruses, most other diseases do not seem to be a problem. If tubers do not develop shoots, they may be sterile or have crown rot.

Whiteflies, leafminers, aphids, red spider mites, and beet army worms will all attack dahlias. Check the labels of the chemicals you commonly use to control these pests to see if the chemicals are registered for dahlias—many are.

Marketing

Ship when the first flowers have started to open. Dahlias can be held for a time at the 55 F temperature mentioned earlier. Removing the terminal flowers and allowing lateral shoots to develop will also slow plants down, but will add an additional five to 10 days to crop time.

99

Teresa Aimone, former editor of GrowerTalks *magazine, is with Sluis and Groot, Fort Wayne, Indiana.*

Culture notes

March 1991

by Sherri Neal

Royal Dahlietta®
Family: Compositae
Genus, species: *Dahlia variabilis*

Dahlietta is, in fact, a dahlia with unique characteristics not found in standard bedding types. The name dahlietta was given to this new type of pot plant to differentiate it from the bedding plant dahlia. Specifically designed for the pot plant producer, dahlietta is versatile and hardy. The short crop time, six to eight weeks from transplanting, allows quick turnover and the opportunity to grow several crops in the time and space normally allotted for a single crop. Dahliettas also make an excellent filler crop between larger scheduled crops.

Dahlietta's dwarf habit and fully double flowers make it an excellent gift plant to present in foil wrap or to display in baskets indoors. On the average, blooms will last nine days, and in warm weather the dahlietta will continue to flower if moved to the patio or into the garden. Dahlietta is as distinct in plant habit from the dahlia as the vegetatively propagated geranium is from a seed geranium.

Dahlia

Propagation
Dahliettas' vegetatively propagated varieties are patented. Virus-free plugs are available in a 98 tray from several suppliers.

Growing on
- **Transplant:** Transplant single plugs into a 4½- to 6-inch pot or three plugs into a 7- or 8-inch pot, using a standard potting mix with a pH of 6 to 6.5.
- **Light:** Flowering response depends on region and growing conditions. Plants need between nine and 13 hours of daylight, depending on light intensity. A minimum 2,000 footcandles is recommended.

Maintaining high light intensity produces a more compact plant. Light intensity also affects production period (number of weeks from transplant to sales availability): six to eight weeks with lower light intensity; five to six weeks with higher light intensity. Under the proper conditions, dahlietta can be grown year-round.
- **Temperature:** Optimum growing temperature is 65 to 70 F days and 60 to 65 F at night. Higher temperatures may result in looser plant habit and will speed up growth.
- **Fertilizer:** Maintain a proper feeding regimen to keep dahlietta in excellent health and create an attractive plant. A weekly feeding of 20-10-20 at 250 to 350 parts per million is adequate.
- **Height control:** Pinch plant one week after transplant to promote branching for added fullness. Dahlietta was bred to be compact; maintaining light requirements eliminates the need for growth regulators.
- **Shipping:** For optimum results, produce dahlietta for the local market. Trials for cross-country shipping indicate positive results for plants shipped in a refrigerated truck at 45 to 48 F.

- **General care:** Dahliettas are light- and water-loving plants. Place them in sunny locations and keep them moist.

Varieties

Dahliettas are currently available in the Royal series of six separate colors. First introduced in 1988, the scarlet, orange, yellow and white colors received awards at the Society of American Florists' New Varieties Competition. Apricot and violet are the most recent additions. These varieties are described in the order of flowering response:
- **Violet.** Compact habit with vibrant double flowers; earliest to flower.
- **Apricot.** Uniform plant habit with plentiful flowers.
- **Yellow.** Vigorous producer with masses of flowers.
- **Orange.** Consistent producer with branching habit.
- **Scarlet.** Vibrant color and strong outdoor performance.
- **White.** Vigorous producer with large double flowers and full plant habit.

Look for more colors in this Royal series in the future. For further information contact: Royal Sluis Inc., 1293 Harkins Road, Salinas, California 93901; (408) 757-4191, FAX (408) 757-5012.

"

Sherri Neal is produce marketing coordinator for the ornamental division of Royal Sluis Inc., Salinas, California.

Delphinium

Culture notes

August 1990

by E. Jay Holcomb and David J. Beattie

Potted delphiniums
Family: Ranunculaceae
Genus, species: *Delphinium* sp.

Delphiniums are garden perennials whose flower spikes have graced perennial borders for many years. One of the advantages of perennials in pots is that they can be dual use plants: They can be enjoyed as a flowering plant in the home, then planted to the garden and enjoyed for years to come.

A tall plant like delphinium doesn't seem the type of plant to grow in a pot, but in flower they are very showy. There are over 300 species of delphinium; some of the modern hybrids are well adapted to pot culture.

Delphinium

In the garden delphiniums range in height from 3 to 8 feet. When choosing a variety for pots, choose one that is naturally short, 2 to 3 feet. The Pacific giant series generally are taller than 4 feet and thus are too tall. The Mid-Century hybrids are beautiful, but are also too tall.

There are a group of varieties that are naturally 2½ to 3 feet, and these are the ones that we have been working with. The variety Blue Springs has worked well in our trials. Blue Fountain and Connecticut Yankee are two other cultivars that should work well.

Propagation

Delphinium can be propagated by division in the garden and also by taking cuttings in spring. This procedure could work for potted plants, but the cuttings supply would severely limit the usefulness of this procedure.

The most practical method of propagating delphinium is from seed. Timing seed sowing depends on when delphinium is to flower. For example, if you were to sow seeds in June, you would have large enough plants to force by fall. Delphinium seed germinates very well when fresh, but store the seed cold after collection.

Sow seed in plug trays in a well-drained medium, covering lightly. Most peatlite mixes have adequate drainage. It's best not to use a plug tray with more than 200 cells so that seedlings have some space to grow before transplanting.

Temperature is critical. Keep medium temperature 70 F with aerial day temperature between 70 and 80 F. Under these conditions germination should occur in 10 to 14 days. When germinating in summer, be careful temperatures don't get too high.

Growing on

Once seed has germinated, grow seedlings on at lower temperatures, as low as 55 to 60 F at night and up to 75 F during the day. For summer the best quality delphinium plugs will be produced in a fan and pad cooled greenhouse. You can also control high day temperatures with shading, but this also reduces light intensity.

Fertilize seedlings lightly. A general purpose fertilizer such as 20-20-20 is suitable, although a fertilizer that's lower in phosphorus and ammonium nitrogen like 15-5-15 produces even more compact seedlings. It takes 7 to 9 weeks for seedlings to grow to a transplantable size of 1 to 2 inches.

It's also possible to purchase plugs from a commercial grower. Some bareroot, field-grown transplants are also available but don't transplant well due to a variety of root-borne diseases.

- **Forcing plants for flowering.** For forcing, grow plants in 5- or 6-inch pots. A 4-inch pot is too small to hold the plant upright. Begin with plugs either from a commercial source or that you grew.
- **Media and fertilizer.** Use a porous, well-drained growing media with a neutral pH. The coarse peatlite mixes work very well. The fertility of the mix should initially be low, but begin additional fertilization soon after transplanting. A complete fertilizer like 20-10-20 at 100 to 200 parts per million at each irrigation is very effective.
- **Light.** Light is critical for delphinium. High irradiance appears to be more important than photoperiod. For example, in November, December and January, the naturally low light levels in Pennsylvania won't stimulate rapid flowering. If you want to flower delphinium in winter use supplementary light.

We have successfully used high intensity discharge lamps at 650 footcandles with the lamps being on from 5 p.m. to 8 a.m. There are probably other supplementary light regimes that will effectively speed up flowering.

In the brighter months of March and April, delphinium grow and flower rapidly under natural greenhouse conditions. We have also flowered delphinium as a cut flower under totally artificial light in about 60 days using large plugs as starting material.

- **Temperature.** Plants tolerate night temperatures as low as 40 to 45 F, but growth rate is much slower. We grow our delphinium at a night temperature of 60 F and a day temperature of 70 F.

In order to keep stem length as short as possible, keep day and night temperatures as close to the same as possible. If temperatures are 60 F night and day, crop time will be a little longer than if the night is 60 F and the day is 70 F.

- **Height control.** Use growing temperatures to control height. There are no growth retardants specifically labelled for delphinium. Our research has shown, however, that some of the retardants like A-Rest and Sumagic do reduce plant height.
- **Pests.** Delphinium will be affected by the same insects that affect other greenhouse crops. Scouting and yellow sticky cards are the first line of defense to controlling insects. Delphinium are also susceptible to some diseases, but starting with a pathogen-free growing media should mean relatively few disease problems.

99

E. Jay Holcomb and David J. Beattie are professors at Penn State University, University Park, Pennsylvania. This research was funded by Bedding Plants Foundation Inc., Lansing, Michigan.

Euphorbia, white top

Culture notes

January 1989

by Teresa Aimone

White Top
Family: Euphorbiaceae
Genus, species: *Euphorbia marginata*

This member of the poinsettia family can be grown as cut foliage or as a cut flower. Also called *Euphorbia variegata*, white top looks especially nice in mixed bouquets.

In areas where summers are cold and wet, it should be grown indoors. White top is early to bloom, and flowers easily midsummer through autumn.

Germinate at 65 to 70 F. Sowing can be done directly into the field in rows, or seed can be sown in trays and then transplanted. Final spacing should be 15

Euphorbia

to 20 plants per square foot. If plants are going to be pinched, plant 10 to 15 plants per square foot. Pinching will prolong crop time slightly, but more stems will be produced.

Optimal growing-on temperature is 60 to 68 F. Some type of stem support system may be necessary. Plant in full sun. Euphorbia can be utilized well in hot, dry areas with poor soils. For best quality, keep plants dry and "hungry."

When stems reach their mature lengths, water intensively to bring out final variegation on upper bracts. Stems will be ready to harvest within one to two weeks.

After harvest, stems should be placed immediately in hot water to prevent loss of the stem's inner latex. Crop time is three to four months.

"

Teresa Aimone, former editor of GrowerTalks *magazine, is with Sluis and Groot, Fort Wayne, Indiana.*

Euphorbia, poinsettia

How to schedule and grow poinsettias for profit

May 1990

by Harry K. Tayama, C.C. Powell and Richard K. Lindquist

Poinsettias are the No. 1 selling pot crop in the United States, but production far exceeds demand and some say quality is at an all-time low. The situation is not hopeless. Read what three experts in the field have to say about poinsettias and marketing, quality and cashing in on the Thanksgiving sales craze.

Marketing

There is a very clear difference between marketing and selling. In simple terms, marketing involves determining consumer preference for products and potential demand for the products. After a producer obtains this information, he manufactures the products. Selling involves manufacturing products and then peddling them as best as possible. Unfortunately, the vast majority of poinsettia growers **sell** rather than **market**.

Euphorbia

Poinsettia production schedules for Thanksgiving sales

Pixie 4½-inch pot, one plant, pinched — Date, 1990

Take and stick unrooted cutting directly in 4½-inch pot	August 6
Cycocel spray at 1,500 parts per million	August 20
Pinch to 5 or 6 leaves if roots at sides and bottom of pot	August 31 (approximate)
Start short days	September 14
Spray with Cycocel (1,500 ppm)/B-Nine (2,500 ppm) tank mix if and when shoots are 2½ inches long	September 17 (approximate)
Stop black cloth shading	October 3
Cycocel spray at 1,500 ppm when effectiveness of Cycocel/B-Nine tank mix spray diminishes, every week, if necessary, until	October 8

6½-inch pot, one plant, pinched — Date, 1990

Take and stick unrooted cutting	July 13
Plant rooted cutting	August 7
Cycocel spray at 1,500 ppm	August 10
Pinch to 7 to 8 leaves if roots are at sides and bottom of pot	August 21 (approximate)
Spray with Cycocel (1,500 ppm)/B-Nine (2,500 ppm) tank mix if and when shoots are 2½ inches long	September 7 (approximate)
Start short days	September 14
Stop black cloth shading	October 3
Cycocel spray at 1,500 ppm when effectiveness of Cycocel/B-Nine tank mix spray diminishes, every week, if necessary, until	October 8

7- to 8-inch pot, two-plants, pinched — Date, 1990

Take and stick unrooted cutting	July 9
Plant rooted cutting	August 3
Cycocel spray at 1,500 ppm	August 6
Pinch to 7 to 8 leaves if and when roots are at sides and bottom of pot	August 17 (approximate)
Spray with Cycocel (1,500ppm)/B-Nine (2,500 ppm) tank mix if and when shoots are 2½ inches long	September 3 (approximate)
Start short days	September 14
Stop black cloth shading	October 3
Cycocel spray at 1,500 ppm when effectiveness of Cycocel/B-Nine tank mix spray diminishes, every week, if necessary, until	October 8

This fact has resulted in the poinsettia becoming a weed in the United States. The poinsettia has become such a cheap crop that a significant number of growers have decided to reduce production in 1990, especially of the one plant pinched, 6½-inch pot. The wholesale price of the one plant pinched, 6½-inch poinsettia has become ridiculously low in certain parts of the United States.

How low? A very prominent grower in Ohio stood before the members in his area grower's organization after the 1989 season and stated, "Let's get the professors at Ohio State University to come to our next meeting and teach us how to grow one plant pinched, 6½-inch poinsettias, sell them for $2.50 each, and make money like some of you think you did this past year."

Euphorbia

Chemicals for whitefly control		
Acephate	PT 1300	5 to 10 seconds per 100 square feet
Bifenthrin	Talstar 10 WP	6 to 24 ounces per 100 gallons
Cyfluthrin	Tempo 2EC	1 to 2 oz. per 100 gal.
Dichlorvos	Vapona, DDVP	Fog or smoke generator
Endosulfan	Thiodan 3EC	21 oz. per 100 gal.
	Thiodan 50 WP	16 oz. per 100 gal.
	Thiodan Smoke	Smoke generator
Fluvalinate	Mavrik Aquaflow	2 to 5 oz. per 100 gal.
Spray Oil	Safer's Sunspray	1 to 2 gal. per 100 gal.
Insecticidal Soap	e.g. Safer's	1.5 to 2 gal. per 100 gal.
Naled	Dibrom	Fog, smoke generator or vaporize off hot surfaces
Neem	Margosan-O	5 pints per 100 gal.
Oxamyl	Oxamyl IOG	1.4 to 1.9 pounds per 1,000 sq. ft.
	Vydate L	16 to 64 ounces per 100 gal.
Resmethrin	SBP-1382	16 oz. per 100 gal.
	PT 1200	5 to 10 sec. per 100 sq. ft.
Sulfotepp	Dithio, Plantfume 103	Fog or smoke generator

If more poinsettias are produced in 1990 than in 1989, what will the wholesale price be? U.S. poinsettia production is out of control, and supply far exceeds demand, particularly for the one plant pinched, 6½-inch.

What can be done to correct or at least ease the situation? Growers can begin by surveying their customers to determine what consumers prefer. They will probably learn that many consumers prefer something other than a one plant pinched, 6½-inch. In 1989, as in past years, grower after grower related that their supply of large sized poinsettias were exhausted early, and they dumped more of the one plant, pinched 6½-inch poinsettias than they cared to admit. For success, poinsettia growers must become involved in marketing instead of selling.

Quality

There is no question in my mind that the quality of poinsettias we are producing today is no better, and perhaps worse, than what was produced 15 years ago, prior to the energy crisis. Poinsettias today are grown closer together and many have hanging baskets overhead. This was not the case before the energy crisis. The result—poorer quality.

Another factor that results in poorer quality is the length of time poinsettias are sleeved, in boxes and transported compared to 15 years ago. It is, of course, easy for a person in my position to say to growers, "Space your poinsettias more, do not place hanging baskets over the poinsettias and take more care in shipping." The question remains, will your customers pay you for the improved quality or buy from another grower who will sell an inferior quality crop at a lower price?

Euphorbia

Thanksgiving

With each passing year, demand for poinsettias during Thanksgiving week has increased. Shopping malls, office buildings, restaurants and department stores are using poinsettias as a prominent part of their decorations. Most poinsettia deliveries are required a few days prior to or immediately after Thanksgiving.

When growers provide poinsettias for Thanksgiving sales, it is always imperative that quality is as outstanding as possible. Poinsettias that are not fully developed when they are delivered most likely will never develop to full potential in a shopping mall, a restaurant or a similar environment.

The following schedules have been developed to have poinsettias—produced in a Northern climate—in prime condition on November 20, 1990, two days prior to Thanksgiving. For Southern production, the entire schedule can probably be delayed by about five to seven days.

Minimum night air temperatures during propagation must be no lower than 70 F. Rooting medium temperatures must be at least 75 F; 80 F is best. After planting rooted cuttings, maintain night air temperatures at 67 F. Whenever possible, keep day temperatures relatively low (less than 67 F) to control height and reduce dependence on growth retardants.

After the appearance of pollen, reduce night temperature gradually until shipping. This will increase bract color intensity and hardens plants so they ship better. As for varieties, grow the ones you know through experience are very tough. Please remember that varieties such as Lilo and V-14 require less growth retardant than the Heggs.

Keep crops good-looking

Poinsettia placement in shopping malls, office buildings, restaurants or department stores is usually performed by interior plantscapers. As a grower, you must communicate with them that bract color will fade in the warm environment of an indoor setting, and this, coupled with damage from irrigating and people, will necessitate replacement in approximately three weeks or by December 11.

In other words, you owe it to the interior plantscaper to package the installation as a two-time poinsettia planting job. Just think of the thousands and thousands of people who will be viewing the poinsettias over a six-week period. We must present our very best appearance and do the best public relations job possible.

Finished poinsettias must be in the best of health (disease free), void of insects and possess long keeping quality. From a health standpoint, the four main diseases that could cause problems are Erwinia, Botrytis, Pythium and Rhizoctonia.

Erwinia (soft rot) can be a serious problem when propagating very soft cuttings under relatively warm conditions. There is no known chemical control for this bacterial pathogen, so strict sanitation must be practiced.

Botrytis can best be controlled by proper heating and ventilation to avoid 100 percent relative humidity. Heating and ventilating at the same time, when switching from day to night temperature to drive out moist air, and maintaining good air circulation in the greenhouse, will go a long way to reduce Botrytis incidence. There are, of course, several chemicals such as Daconil 2787, Benlate DF, Cleary's 3336, Zyban, Ornalin, Phaltan, Chipco 2787, Manzate and Exotherm Termil that control Botrytis.

Pythium is controlled by Truban, Terrazole, Banol or Subdue. **Rhizoctonia** is controlled by Benlate DF or Cleary's 3336.

Euphorbia

Whiteflies (greenhouse and sweetpotato) are becoming more and more difficult to control. For Thanksgiving poinsettias, a good hard freeze may not occur prior to shipping, making control and assurance of whitefly-free poinsettias even more of a problem, since the insect will continue to enter the greenhouse from outdoors. Suggested control chemicals are listed below.

To ensure producing plants with the longest postproduction quality, discontinue fertilization during the last two weeks of crop time.

"

Harry K. Tayama, C.C. Powell and Richard K. Lindquist are professors with the Ohio State University, Columbus, Ohio.

Culture notes

April 1989

by Carolyn Mack

Poinsettia trees
Family: Euphorbiaceae
Genus, species: *Euphorbia pulcherrima*

Poinsettia trees may be grown in a range of sizes, depending primarily on the date the tree is started. The Paul Ecke Poinsettia Ranch in Encinitas, California, produces a trio of trees in three distinct sizes. The largest size is marketed as the Grande Tree and a midsize as the Mini Tree.

The following schedules are just general guidelines. Growers will need to work out their own production schedules depending on their local climatic conditions and past experiences.

Gutbier V-14 Glory Grande Trees

Around May 1, plant an established, rooted cutting into an 8- or 10-inch pot. To ensure height, grow under warm and humid conditions until the first pinch. Stake the plants from the time of potting to help prevent a crooked stem, loosely tying the plants to the stakes as they elongate.

Since trees are grown through summer, heat stress can be a problem. High light intensities can also restrict stem elongation, so grow the plants in a shaded greenhouse (2,500 to 3,000 footcandles) until after the second pinch.

As the plant elongates, remove the very lowest side shoots when they become 1 inch long. Be sure to leave the top 10 shoots to form the branches.

Two pinchings are needed for a desirable shape. The first pinch should be done about August 10. With a regular soft pinch, remove one-half inch of the terminal growing tip. Leave 10 side shoots at the top of the plant, but remove lower remaining shoots.

Three weeks after the first pinch, remove all foliage below the branches. Use a sharp knife to cut off the leaves so that the stem will not be injured.

Euphorbia

The Gutbier V-14 poinsettia Grande Tree (left) and the Mini Tree (right).

Do the second pinch around September 10, leaving two to four nodes on each branch. To shape the plant, leave two to three mature leaves on upper branches and three to four leaves on lower branches. Remove lower leaves slowly as they turn yellow from age.

While the trees require shade during summer, they need full, bright sun after the second pinch. By October 1, remove greenhouse shading.

Adequate spacing during production is essential for a well-shaped finished product. Space plants when leaves begin to touch, and continue to space throughout the growth cycle. Finish Grande Trees at a spacing of 8 square feet per pot.

Watch salt and phosphorus levels in the soil. Since the Grande Tree is in the same soil for seven months, high salts (above 4 EC) may accumulate and phosphorus may be deleted.

Carefully remove stakes after mid-October, once the stem is strong and sturdy, or leave them in place until after delivery to give trees extra support and protection during shipping. Tie up the head of the tree with green colored string to give branches more support.

Cultural requirements from September through December are the same as for any Gutbier V-14 poinsettia.

Trees should be in full flower and ready for market around December 10. Final Grande Tree height is 40 to 48 inches (including the pot) with 25 to 30 blooms.

Gutbier V-14 Glory Mini Trees

Around June 10, pot established, rooted cuttings into a 7-inch azalea pot. Follow the two-pinch program for producing a Grande Tree. Leave six to eight side shoots at the top of the plant; remove lower shoots.

Tone up the trees and give them a nice, compact form with a Cycocel drench at 1:40 or Cycocel spray at 1,500 ppm until October 5 through 10. Final spacing

Euphorbia

should be 7 square feet per pot. Mini Tree finished height is 25 to 30 inches, including the pot.

Shipping and handling care
Poinsettia trees are beautiful but fragile plants that need to be protected and handled carefully during shipping. Sleeving and boxing is highly recommended. Sleeves should have sufficient flair and length to completely enclose the top of the tree to prevent bruising and break ing bracts and branches.

Place sleeved trees in boxes for shipping, with indicator arrows marking which way is up. Use special stakes and inserts to prevent trees from moving in the box. Pack one Grande Tree per box, and four Mini Trees per box.

"

Carolyn Mack is public relations director, Paul Ecke Poinsettias, Encinitas, California.

Growing ideas

August 1988

by P. Allen Hammer

"It's important to be conservative in the use of Bonzi and Sumagic."

Bonzi and Sumagic are excellent height control tools, offering the ability to tailor plants for a desired height with only one to two applications. Many growers have experienced over-application problems with Bonzi and are apprehensive in using such chemicals. However, I suggest giving them another try on a small part of a poinsettia crop.

It's important to be conservative in the use of Bonzi and Sumagic until you fully understand the chemicals. My list of dos and don'ts for poinsettias include: 1) Do not apply after the start of short days. 2) Do not apply in high concentrations. 3) Do not treat all cultivars alike. 4) Apply at the rate of one-half gallon of spray solution per 100 square feet of bench area. 5) Apply spray uniformly to the crop.

My comments are conservative. In research we successfully applied the chemicals in very low dosages to some cultivars after the start of short days without problems. However in other cases, we have also seen severe flowering delay and bract development problems at moderate rates of the chemicals applied after the start of short days. With the control obtained with these chemicals, there should be little need for late application of the chemicals, so why risk a problem?

In 1987, we conducted studies with Bonzi and Sumagic at varying concentrations. Remember, these studies were conducted in the Midwest. More chemical may be needed in the South to achieve equivalent results.

Treatment (parts per million)	Number of Applications	Cultivar					
		Dark Red	Glory	Lady	Brilliant Diamond	V-14 White	V-14 Hot Pink
		Height in inches (percent of control)					
Control		13.1	11.3	11.8	12.5	11.0	10.5
BONZI							
15 ppm	1X	11.4(87%)	10.2(90%)	10.0(85%)	11.4(91%)	10.5(95%)	9.8(93%)
	2X	10.5(80)	9.7(86)	6.4(54)	10.5(84)	10.2(93)	9.3(89)
SUMAGIC							
4 ppm	1X	7.9(60)	10.0(88)	8.4(71)	9.6(77)	9.8(89)	11.0(105)
	2X	8.3(63)	8.8(78)	6.8(58)	6.8(54)	9.1(83)	9.4(90)
8 ppm	1X	9.7(74)	9.4(83)	7.4(63)	7.8(62)	8.3(75)	8.1(77)
	2X	7.2(55)	7.2(64)	6.4(54)	6.2(50)	8.1(74)	7.8(74)
16 ppm	1X	6.8(52)	8.8(78)	6.8(58)	5.5(44)	8.5(77)	7.8(74)

Table 1. Various poinsettia cultivars grown as three single stem plants in a 6-inch pot and treated with various rates of Bonzi and Sumagic. Plants were propagated on August 26, panned September 28, and lighted from September 10 until October 14. First spray: October 7. Second spray: October 14. Mean of three replications (X). Height measured from the top of the pot to upper most bract.

Treatment (parts per million)	Number of Applications	Cultivar					
		Dark Red	Glory	Lady	Brilliant Diamond	V-14 White	V-14 Hot Pink
		Height in inches (percent of control)					
Control		15.4	12.8	13.8	13.5	11.4	11.4
BONZI							
15 ppm	1X	13.9(90%)	11.3(88%)	11.7(80%)	11.0(81%)	10.5(92%)	9.9(87%)
	2X	11.7(76)	11.6(91)	8.8(64)	11.9(88)	9.2(81)	10.3(90)
	3X	10.5(68)	10.9(85)	9.6(70)	9.6(71)	9.9(87)	9.8(86)
30 ppm	1X	12.8(83)	9.8(77)	9.7(70)	10.2(76)	11.0(96)	10.2(89)
	2X	10.5(68)	10.0(78)	8.7(63)	9.8(73)	9.2(81)	9.4(82)
	3X	9.3(60)	9.7(76)	7.5(54)	8.7(64)	8.6(75)	8.8(77)
60 ppm	1X	8.3(54)	9.0(70)				
	2X	7.4(48)	9.0(70)				
	3X	7.1(46)	8.2(64)				
SUMAGIC							
2 ppm	1X	14.3(93)	12.4(97)				
	2X	12.6(82)	10.5(82)				
	3X	11.3(73)	10.5(82)				
4 ppm	1X	12.7(82)	12.0(94)	9.8(71)	11.3(84)	11.3(95)	10.8(95)
	2X	11.1(72)	10.2(80)	8.3(60)	9.3(69)	9.9(87)	9.7(85)
	3X	9.5(62)	9.6(75)	6.8(49)	8.6(64)	8.9(78)	8.8(77)
8 ppm	1X	9.5(62)	11.1(87)	7.7(56)	11.2(83)	10.5(92)	9.8(86)
	2X	7.9(51)	9.1(71)	6.9(50)	8.9(66)	9.6(84)	8.2(72)
	3X	7.8(51)	8.3(65)	5.4(39)	6.3(47)	8.1(71)	7.9(69)
16 ppm	1X	6.5(42)	9.3(73)				
	2X	6.2(40)	7.9(62)				
	3X	6.0(39)	7.6(59)				

Table 2. Various poinsettia cultivars grown as single pinched plants in a 6-inch pot and treated with various rates of Bonzi and Sumagic. Plants were propagated August 3, panned August 31, pinched September 11, and lighted from September 10 until October 9. First spray: September 18. Second spray: September 25. Third spray: October 2. Mean of five replications (X). Height measured from the top of the pot to upper most bract.

Euphorbia

		Cultivar	
Treatment (parts per million)	Date of Application	Dark Red	Glory
		Height in inches (percent of control)	
Control		15.2	12.9
BONZI			
15 ppm	September 2	13.9 (91%)	12.5(97%)
30 ppm	September 2	11.5(76)	12.6(98)
60 ppm	September 2	10.6(70)	10.7(83)
SUMAGIC			
2 ppm	September 2	15.7(103)	12.6(98)
4 ppm	September 2	12.8(84)	12.3(95)
8 ppm	September 2	10.8(71)	12.0(93)
16 ppm	September 2	9.3(61)	10.8(84)
8 ppm	September 9	10.3(68)	10.3(80)
16 ppm	September 9	10.0(66)	10.6(82)

Table 3. Poinsettia Dark Red and Glory grown as single pinched plants in a 6-inch pot and treated with Bonzi and Sumagic one time near or at the start of short days. Plants were propagated August 3, panned August 31, pinched September 11, and lighted from September 10 until October 9. Mean of five replications (X). Height measured from the top of the pot to the upper most bract.

The main points in the data: 1) The Hegg type cultivars are more responsive to both chemicals than Glory. 2) Multiple applications inhibit stem elongation. 3) One spray application made near the start of short days gave similar final height controls as the same application made earlier in the production cycle.

"

P. Allen Hammer is professor of floriculture, Purdue University, West Lafayette, Indiana.

Thanks to Terri Kirk, research technician, Purdue University, for help in conducting these research studies, Paul Ecke Poinsettias for supplying the plants for study and Chevron Chemical Co. for financial support of this study.

Euphorbia

Regional evaluations of new poinsettia varieties

by C. Anne Whealy

New poinsettia varieties are more plentiful than ever. We asked researchers from Ohio, North Carolina and Florida for their evaluations of the new varieties that they have grown. Here are their comments. Remarks from Ohio also include observations from industry personnel. Poinsettias in the Bradenton, Florida trials were grown in polypropylene shade cloth structures as well as greenhouses. These observations are intended to be used as guidelines for poinsettia producers in their selection of new varieties for their area.

Researchers and location	Eckespoint Lilo Bright ruby red	Eckespoint Celebrate (C-27) Light bright red	Gross Supjibi Bright red	Regal Velvet (H518) Dark red	Noel Soft red	H365 Dark red
Dr. Harry Tayama The Ohio State University Columbus	Good branching Good dark bract and foliage color, striking Nice positioning, large number and size of bracts Should be grown pinched, two cuttings per 6-inch or one per 5-inch Good keeping quality	Evaluation not available	Variety not evaluated	Good branching Good bract color and plant form Held color well Not as nice as V-14 Glory	Early flowering Good bract color and plant form Held color well Not as nice as V-14 Glory	As good as Annette Hegg Dark Red Large, vigorous plants
Dr. Roy Larson Beth Thorne North Carolina State University Raleigh	Good as a 5-inch pinched, one cutting per pot No epinasty Best keeping quality of all varieties tested	Seven breaks per pinched plant Large bract size Spectacular Attractive as single-stem or pinched No epinasty Good keeping quality	Freely breaking Recommend growing as a pinched plant Very wide bracts No epinasty	Variety not evaluated	Variety not evaluated	Variety not evaluated
Dr. Jim Barrett Dr. Terril Nell University of Florida Gainesville	Poor branching Good leaf color Good quality single-stem plant with two to three cuttings per 6-inch pot No epinasty Excellent post-production quality	Good branching Looked promising Good potential for southern growers, warrants further evaluation	Good branching Good plant structure Good bract color Good potential for southern growers, warrants further evaluation	Poor branching Good quality as single-stem, 2 to 3 cuttings per 6-inch pot Dark leaf and bract color	Similar to a Hegg Not as nice as H365	Similar to Annette Hegg Dark Red in leaf color and plant structure Fairly good stem strength Good bract color Good quality plant
Dr. Gary Wilfret Gulf Coast Research & Education Center Bradenton, Florida	Poor breaking with high temperatures Flowered late Bracts had a smokey tinge	Branching not as good as V-14 Glory, but better than Lilo Good bract color Bracts held in a good display	Good branching Very early Many bracts Foliage keeping quality poorer than V-14 Glory, but better than V-10 Amy	Erratic formation of laterals like Lilo Strong stems Good bract color	Good branching Good stem strength Small bract size Average keeping quality	Early flowering Large bracts Good keeping quality Greatest potential of all new reds for Sunbelt producers

—continued on next page—

Euphorbia

Researchers and location	Merrimaker Dark red	Minstrel Red	Frost Creamy white	Mirabelle White	Minnekin Pink	Eckespoint Jingle Bells 3 Dark red and pink novelty	Noel Blush Light pink and white novelty
Dr. Harry Tayama The Ohio State University Columbus	Variety not evaluated	Variety not evaluated	Early flowering Not a pure white like Annette Hegg Top White A satisfactory substitute for Top White, but no real improvement	Variety not evaluated	Variety not evaluated	Variety not evaluated	Variety not evaluated
Dr. Roy Larson Beth Thorne North Carolina State University Raleigh	Vigorous grower Magnificent red Vivid floral display	Early blooming Doesn't require black cloth	Variety not evaluated	Variety not evaluated	Variety not evaluated	Variety not evaluated	Variety not evaluated
Dr. Jim Barrett Dr. Terril Nell University of Florida Gainesville	Poor plant form, spreading habit	Early flowering, black cloth not required Good color Very dwarf, probably only good in a 4- or 5-inch pot, too small for a 6-inch	Variety not evaluated	Variety not evaluated	Variety not evaluated	Variety not evaluated	Variety not evaluated
Dr. Gary Wilfret Gulf Coast Research & Education Center Bradenton, Florida	Too vigorous Strong, thick stems Laterals angle outward causing open center, poor form Poor postproduction quality	Good for 4-inch pots Too small for anything over a 4½-inch Early flowering	Good stem strength More cream colored, not a true white like Annette Hegg Top White Poor postproduction quality	Should only be produced as a 4-inch Not enough bracts Poor keeping quality	Should only be produced as a 4-inch Not a good pink, Too fleshy	Good branching One week to 10 days later than Glory Good bract color clarity between red and rose	Best novelty Strong stems Good bract keeping quality and average foliage keeping quality Good possibilities for southern growers

Eckespoint Lilo, Eckespoint Celebrate, Gross Supjibi and Eckespoint Jingle Bells 3 are available from the Paul Ecke Ranch. Regal Velvet, Noel, H365, Frost and Noel Blush are available from Ball Seed. Merrimaker, Minstrel, Mirabelle and Minnekin are available from Mikkelsen's.

Eustoma, Lisianthus

Culture notes

July 1990

by Daniel J. Jacques

Blue Lisa eustoma
Family: Gentianaceae
Genus, species: *Eustoma grandiflorum*
Common names: Lisianthus, Prairie gentian

Prairie gentian is described in *Hortus Third* as being native to the southwestern United States east to Florida and south to Mexico, the West Indies and the northern portions of South America. The cultivated form was once classified as *Lisianthus russellianus* but is now classified as *Eustoma grandiflorum*.

Early cultivated varieties of eustoma were commonly grown as a cut flower crop. Plants required pinching and growth regulators to produce an acceptable pot crop. The major markets for these flowers have been Japan and Holland.

Recent breeding efforts have been directed toward obtaining compact varieties with better branching. One variety, Blue Lisa, is a 1990 FloraStar winner for outstanding pot performance, producing deep blue flowers on a compact plant. It's a variety that requires no growth regulators or pinching in order to produce a suitable pot plant. It can be produced in either 4-inch (one plant per pot) or 6-inch pots (three plants per pot) and can be used in the landscape, tolerating hot, dry weather after becoming established.

Propagation
Sow seed in a good germinating mix in plug trays. We recommend the 400-size plug trays; plants seem to do best in these. Seed require light for germination, so don't cover them. Germinate at 75 F. Germination will occur in about 10 to 14 days.

Grow plugs at 60 to 68 F; don't allow them to dry out. Feed plants with 75 ppm N constant liquid feed (CLF). An occasional leaching with clear water keeps soluble salts levels down.

Growing on
• **Transplanting.** Shift your plugs to 72-cell trays after the first two pairs of true leaves have developed. At this time, transplant plugs from the 400 trays directly to a 4-inch pot. Use plants from the 72-cell trays either for 4- or 6-inch pots. Don't allow plugs to become root bound in the 400 trays.

Shift plugs to 4- or 6-inch pots. Use plugs from the 72-cell trays for the 6-inch pots. Use plugs from either 400 trays or 72-cell trays for the 4-inch pots. Keep plants from becoming stressed after transplanting as this may reduce branching.

Eustoma

- **Fertilizing.** Fertilize plants with 150 to 200 ppm N, CLF. Use a well-balanced fertilizer formulated for use in soilless media (Peters 15-16-17 or 15-17-17 Peat-Lite works fine). It's also a good idea to occasionally leach the plants with clear water in order to decrease soluble salt levels. Keep plants well watered, but don't overwater.
- **Flowering.** Eustoma needs long days and warm temperatures (60 F nights and 70 F days) in order to flower. They will flower at lower temperatures, but it takes longer.

To flower Blue Lisa in winter, some type of daylength extension is required. We recommend high intensity discharge lights for day extension. Growers wishing to use mum lighting (4-hour nightbreak with incandescent bulbs) may need to pinch the crop.

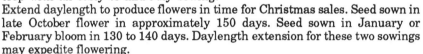

Produce Blue Lisa for a Christmas crop with an early to mid-July sowing. Extend daylength to produce flowers in time for Christmas sales. Seed sown in late October flower in approximately 150 days. Seed sown in January or February bloom in 130 to 140 days. Daylength extension for these two sowings may expedite flowering.

- **Diseases and pests.** Eustoma can be susceptible to Rhizoctonia, Fusarium and Botrytis. It's always important to keep the growing and transplanting areas clear of plant debris to help keep disease occurrence down to a minimum or even prevent diseases from occurring. Using horizontal air flow and avoiding overwatering or watering late in the day also helps to minimize disease occurrences.

If diseases do occur, chemical control is possible. Use Benlate for all three of the diseases mentioned, but use it in rotation with other fungicides. Other fungicides that can be used include Banrot for Rhizoctonia and Fusarium, Terraclor for Rhizoctonia, and Chipco and Daconil for Botrytis. Other chemicals not mentioned here might also be of benefit, but check with your local extension agent to determine if they can be used.

Aphids and thrips can become a problem if not properly controlled. Keep all work areas clean to decrease the possibility of large infestations. Monitoring insect populations is very important.

Mavrik and Avid work for both aphids and thrips. Also control aphids with Orthene or Diazinon. Control thrips with Dycarb or Dursban in addition to Mavrik or Avid. It's best to rotate the chemicals for control.

Blue Lisa eustoma can be a very good landscape crop due to its compact habit and heat and drought tolerance. It's less labor intensive than varieties previously grown for pot production. It will be a very profitable crop—a crop definitely worth trying.

Daniel J. Jacques is with Grace-Sierra, Fogelsville, Pennsylvania.

Culture notes

April 1990

by Jack S. Sweet and Paul Cummiskey

Exacum
Family: Gentianaceae
Genus, species: *Exacum* sp.

This beautiful blue-flowered plant has exploded into popularity as a pot plant in just a few years. The myriads of dime-sized flowers—blue with bright yellow pollen masses in the center—tend to cover the whole plant when grown well. Plant and pot size vary with the grower and his market.

Propagation
Most exacum are purchased from specialty growers as small plants ready for potting into 4-, 5- or 6-inch pots. These plantlets may be from cuttings or seed. Seed is available, but is quite tiny—even smaller than begonia seed—and must be handled very carefully. Exacum seed germinate slowly—2 to 3 weeks. Plant in a lightweight starting media with little or no covering. Transplant seedlings in about 6 weeks into 2-inch pots. Put into final pots for flowering about 7 weeks later.

Culture and production time
The cultural procedures that follow are based on using plants produced by specialty growers ready for final transplant from 1½- to 2-inch pots.
- **Potting.** A light, well-drained potting soil that's loose and has plenty of soil amendments (perlite, calcine clay or styrofoam) and peat allows good root aeration. Place one plant per pot deeply enough in the soil to stabilize the plant as it grows. Water lightly for the first 10 days to encourage roots.
- **Temperature.** Exacum grow best at 60 F to 65 F nights and 75 F to 80 F days.
- **Light.** Grow under full sunlight in winter months. During summer, apply light shade (4,500 to 6,000 footcandles). During spring and early fall, grow in full sun; apply light shade when plants start to flower. Shading produces darker colored blooms; excessive light and heat fade flowers.
- **Fertilization.** Exacum are moderate feeders. Alternate fertilizers such as 15-16-17 Peatlite Special and calcium nitrate at 2 pounds per 100 gallons every third watering. If desired, incorporate Osmocote 14-14-14 in the media at ¼ to ½ teaspoon per 6-inch.

Exacum

Excessive leaf curl or crinkle may be related to excessive light or possibly a low copper level in the leaf. A foliar spray using Tri-Basic Copper at 1 pound per 100 gallons applied 2 weeks after potting reduces crinkle on exacum. Soil applications of copper haven't been too useful.

- **Height control**. Control plant height with B-Nine at regular mum strength (.25 percent solution) applied 1 week after potting and, if needed, a second application 2 to 3 weeks later. Regulating water also controls height. If small plants are desired allow pots to dry out more between waterings.
- **Lighting**. Exacum react to total light energy on leaves. Flower bud initiation isn't affected by daylength, but growth is increased by longer days. Using supplemental lighting in winter is beneficial. Lighting with HID lights or mum-type lighting of 10 to 20 fc 4 to 6 hours a night, starting at dusk, can speed up winter production by 2 or more weeks.
- **Winter growing**. Winter produces more growing problems than summer. Lower light levels and shorter days make a softer plant that is easily injured and attacked by disease. Compensate by using lower fertilizer levels and reduce watering to make a harder plant. Around October, dry pots thoroughly between waterings. Water early in the morning so foliage is dry by late afternoon. Provide good air circulation and reduce fertilizer levels by half. Overwatering and high nutrient levels promote disease and delay flowering. Light at normal mum intensity forces winter plants into bud and bloom.
- **Timing is seasonally variable** with a marketable 6-inch flowering plant in 7 to 8 weeks in summer and 12 to 14 weeks in mid-winter. Plants in $4^{1}/_{2}$- or 5-inch pots for mass market sales can be produced in less time and on less bench space, as exacum can be forced to flower at an early stage. No pinching is necessary; they are self-branching plants. In case of premature budding, remove the earliest flowers to produce larger plants.
- **Disease and insects**. The major disease problem with exacum seems to be related to Botrytis (gray mold). It can show up as a gray lesion at the soil line or small gray lesions at the forks in branches and cause all or part of the plant to wilt and die. This Botrytis is actually caused by excess fertilizer, too much water or water left on the leaves in evening or overnight. Botrytis control chemicals such as Chipco 26019 (Rovral) at 16 ounces per 100 gallons as a foliar spray only, or Benlate at 6 ounces plus Daconil 75 WP at 10 ounces per 100 gallons as a drench or spray helps control this problem, but less fertilizer strength and less water are also needed.

Pythium and Phytophthora can be controlled by a tank mix of Benlate at 8 ounces and Subdue 2E at 1 to 1½ ounces per 100 gallons. Banrot at 8 ounces per 100 gallons is also useful as a light drench right after planting. Regular use of the above chemicals can eliminate most all diseases on exacum, but remember—they need it right after potting and repeated two to three times during production.

The most damaging insect to exacum is the broad mite, usually found on the upper plant parts. It causes leaves and growing tips to become yellow and distorted, and buds will fail to open. Control broad mite with various miticides (Pentac, Kelthane, Vendex). All three cause flower injury so caution is advised on mature flowering plants. Avid 0.15 EC has proven useful in mite control and doesn't injure flowers.

Varieties

There are several varieties used by commercial growers: Blue Champion, White Champion, Royal Blue (a new deeper color that comes in 7 to 10 days earlier), and Little Champ (a smaller F_1 hybrid for the 4- to 4½-inch trade. It's smaller, sky blue and 15 to 20 days earlier than Blue Champion).

99

Jack S. Sweet and Paul Cummiskey are with Earl J. Small Growers Inc., Pinellas, Florida.

Fill out your sales with potted freesia

May 1990

by Miriam Levy

The sweet fragrance of freesia makes them an ideal alternative for holiday sales. With new cultivars and height control chemical dips, potted production has never been easier.

Freesia were introduced to horticulture in the mid-1880s as a fragrant cut flower. They are native to South Africa and are available year-round in a wide range of colors. The past few years, freesia have gained increased popularity as a low temperature pot crop. This popularity resulted in increased research from universities, focusing on reducing days to flower and plant height. As a result, freesia are easier to grow and more growers are entering the freesia market.

Freesia are available in a wide range of colors—blue, lavender, white, yellows and shades of pink to red—and as single or double flowers. They can be produced from seed or corms, which are the most popular. Seed-grown freesia require seven to eight months from seed to flower, while corm production requires less time—four to five months from planting to bloom. The most common planting times are September through December for blooms January through April.

Handling purchased corms

Freesia corms are produced in Holland, where they are treated at 86 F for 13 weeks after harvest to break dormancy. Gus De Hertogh of North Carolina State University recommends air shipments to the United States to avoid long shipping times at high temperatures. Unpack corms and immediately give them a fungicide dip.

George Wulster and Tom Gianfagna of Rutgers University recommend the use of Benlate at the rate of 1 pound per gallon of water for 30 minutes. This dip helps prevent development of Fusarium during the growth cycle. Dry freesia corms overnight at a maximum temperature of 54 F.

Begin cold storage immediately to develop flowers. Store corms in open trays at 42 to 45 F for 35 to 40 days with high relative humidity and good air circulation. Storage over 49 days causes corms to pupate and may decrease plant vigor. Cold treatment reduces flower stem height and time to flower; too

Freesia

Fusarium (left) and virus infections (above) present major problems to potted freesia, destroying foliage and detracting from saleable quality. Control fusarium with a pre-plant corm dip; the trick to virus control is hygiene—start with clean corms and keep virus-transmiting aphids out of the growing area. *Photos courtesy of Jens Jorgen Brondum, Denmark.*

high temperatures during this period either prevent flower development or result in malformed blooms.

Height control for corms

Freesia varieties on the market today are bred for cut flower production. Therefore, the corms should have a special treatment if they are to be used as a potted crop. Rutgers and North Carolina State University and the Institute of Glasshouse Crops in Denmark have researched pre-plant growth regulator dips for height control. Both Bonzi and A-Rest are effective, but Bonzi is one-tenth as expensive as A-Rest, which doesn't yet have EPA registration for use on freesia.

When using Bonzi as a pre-plant dip, apply it at the rate of 100 parts per million for one hour. (3.4 fluid ounces of 0.4 percent active ingredient per gallon of water.) Some varieties may require up to four hours' soak time.

If used as a post planting soil drench, apply when shoots are 1½ to 2 inches in height. Try a drench solution of 12 ppm Bonzi at the rate of 100 milliliters per 6-inch pot, 60 milliliters per 5-inch pot. If Bonzi is used as a dip, plant corms while they're still wet.

Freesia varieties suitable for pot growing

Variety	Flower	Days to flower	Plant height (inches)
Blue Heaven	blue/single	110	12-14
Blue Navy	blue/semi-double	100	14-16
Golden Melody	yellow/single	105	14-16
Jessica	pink/semi-double	100	12-14
Pink Westland	pink/semi-double	100	14-16
Polaris	white/single	110	16-18

Like most bulbs, freesia are fluoride sensitive with damage being expressed as leaf scorch. Do not incorporate perlite and superphosphate into the soil for this reason. A well-drained medium with a pH of 6.5 to 7.0 is ideal. Plant corms 1 inch deep in a standard pot. Most growers plant four to six corms in a 4-inch pot, six to 10 per 6-inch pot, and 10 to 16 in an 8-inch pot. Freesia require medium to high light levels, over 2,500 footcandles for optimal growth and flowering. Low light levels during flower development can cause bud abortion.

Get best growth with low temperatures

As with most bulbs, freesia growth is temperature dependent. Until shoots emerge (about three weeks), keep soil and air temperatures 54 to 55 F. Upon emergence, day temperatures of 65 to 66 F and night temperatures of 58 to 60 F are optimum. Day temperatures above 68 F result in taller plants, excessive foliage growth and flower bud abortion. Day temperatures below 60 F slow down flowering time.

Freesia are not heavy feeders. After growth begins, an application of 100 ppm nitrogen (preferably in the nitrate form) of a balanced fertilizer once a week is sufficient.

If growing conditions are not ideal, freesia tend to get leggy. This is caused by high temperatures during forcing, inadequate cooling or low light levels. Most growers stake their freesia using special rings or bamboo stakes and string. Dutch breeders are working to develop naturally compact varieties that should be available within two to three years.

Virus and Fusarium are the major problems

Botrytis and aphids are two common problems encountered in freesia production. Both can be controlled with standard greenhouse practices.

Virus infections can also be a problem. Attack by virus is typically seen as dark spots on the leaves, which gradually wilt, or as deformities in florets. Some varieties are more susceptible than others and a considerable difference in infection occurs in different seasons. Virus cannot be eliminated by chemicals; only the firm delivering the corms can prevent infection by tools and aphids.

Freesias are also susceptible to attack by Fusarium, which is transmitted by corms. Attacks are seen as wilted leaves and/or non-sprouting corms. Dip corms in Benlate or apply a Benlate drench to prevent Fusarium attacks.

Remove pots showing symptoms at once. To freesias in pot culture, Fusarium and virus are serious diseases, because leaves are an integral part of the sales product.

How to win with pot freesias

Market freesia when the first floret opens. Flowers open consecutively along the spike, lasting six to 10 days. Removing faded flowers encourages new buds to open. Flowering can be delayed or accelerated in the greenhouse by reducing or increasing temperature during the final forcing stage. If freesia must be held, lower greenhouse temperatures. Storage is not really an option, says George Wulster, Rutgers University. Material out of storage initially looks fine, but plants don't recover and flowers won't open.

All research indicates a "hybrid" method of producing dwarf potted freesia. This includes proper cultivar selection, cold treatment of corms before planting, application of growth retardants either as a drench or dip and proper temperature control during the forcing period.

The sweet fragrance of freesia makes them ideal for holiday sales. As interest grows and research on culture continues, pot freesia will find their profitable niche in the marketplace.

"

Miriam Levy is with Sluis and Groot, Capitola, California.

Cut freesias year-round

January 1986

Debbie Hamrick

Fragrant freesia stems are turning up with increasing frequency. Fin floral arrangements, bridal bouquets, and in supermarket flower coolers all across North America. Few cut flowers can match the freesia's fragrance, variety of colors, and longevity. The best part is that consumers want them, and more and more consumers are learning what they are.

At the seventh annual Canadian Greenhouse Conference in Guelph, Ontario, George Ravenek, Ravenek Greenhouses Ltd., Langley, British Columbia, shared his freesia growing techniques.

As evidence of the freesia's rapid rise in popularity, George cited several statistics. Last March at the Vancouver flower auction the volume of freesias sold increased by 350 percent to 400 percent over the same month in 1984. "It was not unusual to see 5,000 to 6,000 bunches of freesias move through on a given day," he says. Overall, freesia sales for the month of March at the auction were 63,000 bunches—that's 630,000 stems! That number climbed to 88,000 bunches in April.

And, out of all the freesias grown throughout British Columbia, George says 80 to 90 percent are sold through the auction. At Ravenek Greenhouses workers plant approximately 20,000 freesia comes monthly, Beds are completely turned over twice during the year. Estimated return on the crop is roughly $8 (Canadian) per square foot.

Freesia

Step by step from the beginning

Soil: George recommends a loamy well-drained soil. The loam makes digging or lifting corms easier, he says. Work the beds to an 8 to 10 inch depth.

To control Fusarium, beds are deep-sterilized for five hours prior to planting. Once the corms are planted, George warns, there's little defense against Fusarium. And, without sterilization, you can't grow very many consecutive crops before running into trouble. Additionally, one wheelbarrow of manure, as well as dolomitic limestone, and superphosphate are incorporated for every 100 square feet. (The *Holland Bulb Forcer's Guide* warns against additives containing fluorides.) Maintain a pH of 6.0 to 6.5.

Spacing: George uses 5-inch by 5-inch square chrysanthemum netting to guide workers when planting; the netting later serves as support for the plants. Crews plant anywhere from one to 1½ to two corms per square depending on the variety and corm size. A second chrysanthemum net is installed right away; freesias need support during growing.

Temperature: "The temperature after planning is so important to the success of the crop," George notes.

For regular-sized corms, maintain 58 F for one month. Cormlets, or smaller corms, should be maintained at 62 F for six weeks. These higher temperatures for cormlets are necessary in order to grow a good-sized plant before bud set, he says.

If soil heating is available during the winter months, George recommends that beds be covered with 2-mil plastic for four weeks. During this period, greenhouse temperatures can be maintained at 46 F. When plants are 6 inches tall, remove the plastic. To guard against burning, make sure to remove the plastic on a cloudy day or late in the afternoon.

Growing on: After the initial four (corms) to six (cormlets) week temperature period, gradually lower the temperature to 53 F. "The temperature can be a little higher on bright days and a bit lower on dark days," he says. During the growing season make sure to raise the netting.

Flowering: Plants should flower in 120 days. By then, the crop will be 2½ to 3 inches tall. Each corm should produce one main stem and anywhere from one to three or more side shoots depending on the variety.

Cutting: "We cut freesias a bit more open than European growers, and I think we have an advantage there. Stems cut when the flowers are more open tend to perform better in the vase," he says.

During winter months, George cuts the crop three times per week with the crop lashing over a 4- to 6-week period. On warm summer days, stems are cut daily and a bit tighter. The summer crop flowers out faster-in three weeks.

Cutting, grading, and bunching (10 stems per bunch) is all done in the greenhouse. Workers cut #1 grades first, then go back for the #2 grades. The #1 grade features long straight 14-inch stems with at least five buds. The #2 grade measures 10 to 12 inches long with at least three buds. Until stems are sold, they are held in the cooler at 37 to 40 F.

Cooling the soil

Eight years ago (and quite by accident), George stumbled across the benefits of soil cooling for summer freesia production. In the freesia houses he noticed one 5-foot-wide strip running east/west where the plants flowered a month earlier. The first year he didn't pay any attention to it, but then he realized that those plants were growing right over the water main from his well.

Based on this observation, George began experimenting with various pipe sizes, spacings, and mulches to more closely control the soil temperature during the warm months. This technique of cooling the soil has become quite popular with growers in both North America and Europe, George says.

Based on the results of his experiments and experience, George now uses a series of 3/4-inch PVC pipes buried 12 to 16 inches deep and spaced 1 foot apart to funnel 1,500 gallons per hour of 48 F well water underneath his freesia beds. The water flows 24 hours a day from April to August.

"It's important," George explains, "that the soil temperature doesn't rise above 64 F or the buds will not set. To maintain cool soil temperatures during warm weather it is also necessary to cover the soil [in addition to cooling the soil." After trying several different mulches, George settled on styrofoam chips. Beds are covered with 2 to 3 inches of chips immediately after planting. Once buds have set, eight to 10 weeks later, the chips are removed with a homemade vacuum system.

To further control temperatures during spring and summer months, George uses a computer-controlled 60 percent shading system. "This way the crop benefits from early morning and late evening light." Shading also helps to control both humidity and corm aging, he adds.

After flowering

Freesia corms may be flowered for five years or more as long as the grower is fussy about virus.

Aging corms: Once flowers are cut, corms are left in the beds for four to six weeks to mature. When mature, they are dug, cleaned, graded, then dried quickly.

Corms to be planted four months later are then stored at 80 F at 87 percent humidity for 3½ months. This heat treatment is necessary to break corm dormancy for re-flowering. After the heat treatment, corms should show some root activity, George says. Do not replant corms showing any sign of virus or disease.

"

Debbie Hamrick is publisher/editor of FloraCulture International *magazine and international editor of* GrowerTalks *magazine.*

Culture Notes

November 1985

by Teresa Aimone

Fuchsia
Family: Onagraceae
Genus, species: *Fuchsia* x *hybrida*

Fuchsias are a sure-sell item for retailers and wholesalers. The flowers consist of a tubular corolla that extends from four spreading sepals and comes in combinations of red, pink, white, and purple. When in full bloom, these cascading beauties sell themselves.

Due to their trailing habit, fuchsias are most commonly produced for basket sales, though they can be produced in pots; or, for the creative customer, they can be produced in packs for a "make your own basket." For a longer-term crop, but with equally spectacular results, fuchsias can be grown in tree forms.

Propagation
Though fuchsias can be grown from seed, it's not a common practice. Take 2-to 3-inch terminal cuttings from stock plants. Fuchsias flower from lateral axes, so stem cuttings will result in more reproductive rather than vegetative growth; therefore, terminal cuttings should be taken. Most varieties will root in 14 to 21 days. Though it's not necessary, the basal end of the cuttings can be dusted with a 1 percent IBA solution. Rooting will occur without the hormone, but it will be faster if it's used.

Maintain a rooting media temperature of 65 to 70 F. Drench the cuttings after planting into the propagation medium. Syringe or mist the cuttings as needed to keep them turgid.

When taking the cuttings, make sure that the cutting material is mature enough—not too herbaceous. Cuttings should "snap" when they are taken. For spring crop sales, take the cuttings prior to March 1. Once the cuttings have rooted, transplant them directly into the final containers.

Growing-on stage
Medium: Fuchsias respond well to soil combinations that provide adequate water-holding capacity, but also have good drainage, adequate cation exchange capacity, and proper aeration. Provide a good organic content in the medium since fuchsias are heavy feeders. If you're growing fuchsias for baskets sales, you may want to use a lighter mix than you would for pots.

Fuchsia

Fuchsias are heavy feeders, and they respond well to additional levels of N, P, and K in the medium—you may wish to incorporate Osmocote or MagAmp into the medium prior to planting.

pH: Maintain a pH level of 6.0 to 6.6 for best results.

Lighting: Fuchsias are long-day plants; when exposed to daylengths longer than 12 hours, the plants will initiate and develop flower buds. Less than 12 hours of daylength and they will remain vegetative. Naturally long days (12 hours or more) occur from March 1 through October 1; so for winter flowering, you need to provide artificially long days.

Provide plants with 10 to 20 footcandles of incandescent light from 10 p.m. to 2 a.m. during the period of October 1 to March 1. Plants will flower 45 days after the start of long days.

If light intensities fall below 450 to 500 footcandles during any time of the year, flowering will be delayed. And without the use of supplemental lighting, plants may be hard to flower in the North. Avoid direct sun on fuchsias—it can burn the foliage.

Temperature: Fuchsias are also very temperature-responsive. At 73 F, flowering will occur after plants have been exposed to 40 to 45 days of long-day treatment. If lower temperatures are used, add approximately one day for every degree lower than the optimum 73 F to calculate the approximate time of flowering. When fuchsias are grown at temperatures below 60 F, flowering will be delayed.

Pinching: Once the plants have been established and are growing, pinch back to the fourth node (six nodes should have developed before this is done).

If the desired selling date is Memorial Day, make the final pinch seven to eight weeks prior to sale. If flowering for Mother's Day is required, allow eight to nine weeks from the last pinch to sale.

Fertilizing: Fuchsias are heavy feeders and like lots of water—especially during the spring and summer months. Water with constant liquid feed of 200 to 250 ppm, 20-20-20; this feed program can be alternated with the same ppm of nitrogen and potassium using calcium nitrate and potassium nitrate.

Insects and diseases: Spider mites are the most common pests of fuchsias, though whiteflies and aphids will also attack fuchsias. Most available chemicals used to control these pests are cleared for fuchsias. Check the label before applying.

Botrytis can be a problem with fuchsias. Be on the lookout for this disease especially in the spring and during cool, cloudy periods. Exotherm Termil can be used for Botrytis control. A monthly preventative drench of Lesan/Benlate is also recommended.

Fuchsia

Fuchsia tree forms

Fuchsias make spectacular tree forms, and production time to produce one of these beauties is approximately one year. To produce a fuchsia tree, stake the plants, and tie it for support as it grows up the stake. Remove all lower shoots, but keep the top branches intact. When the plants have reached their final desired height, pinch to produce a bushy, tree-like head. Pinch any new shoots when they have developed four nodes. Repeat pinches until the desired shape and size is reached.

Gibberellic acid can be used to elongate the stem. GA will not add new cells to the stem, it will stretch existing ones, so the stems will be much weaker than if you had not used GA. Recommendations say to use a 250 ppm solution (one Gib-Tab per four gallons of water). This solution should cover 200 square feet if sprayed to runoff. These same recommendations suggest applying four sprays at weekly intervals to induce tree forms. Be careful when using gibberellic acid. It is a naturally occurring plant hormone and very little is needed to produce a dramatic affect. Closely watch how the plants respond and judge accordingly.

Varieties

Fuchsias come in a wide variety of color combinations and flower forms (single or double). Performance varies per variety. Two good cascading basket varieties that work well in the North are Swingtime (red sepals with white corona) and Dark Eyes (red sepals with purple corona).

Other varieties include: singles—Mrs. Marshall, Glendale, Display, Black and Purple Prince, Pink Beauty, Little Beauty, and Lord Byron (the last three varieties are early—suitable for Mother's Day). Double varieties tend to have a more erect habit and include Giant Double Pink, Pride of Orion, and Gladiator.

Other suitable hanging basket varieties include Inca Maiden, Gay Senorita, Cascade, and Halloween.

"

Teresa Aimone, former editor of GrowerTalks *magazine, is with Sluis and Groot, Fort Wayne, Indiana.*

Gerbera

Culture Notes

June 1987

by Lewis Howe

Transvaal Daisy
Family: Composite
Genus, species: *Gerbera jamesonii*

Gerberas, or transvaal daisies, have gained much attention as a potted plant in the last 10 years. Although Gerberas make excellent cut flowers, breeding advancements have produced new and improved pot types from seed.

Gerberas are also popular bedding plants in the Southern U.S. for borders and mass plantings.

Propagation

Liners: Gerberas have been cloned through tissue culture for many years. Advantages of clonal selections are true flower color and uniform habit and blooming of the entire crop.

Many large producers can purchase ready-to-transplant tissue-cultured plants or ½- to 2½-inch plugs to pot into 4-, 5-, or 6-inch pots. After receiving liners, immediately unpack and pot them in a sterile peat-lite media. Place the soil level of the liner even with pot soil level. Crowns should not be covered. Cultural procedures for liners are the same as those listed for seedlings.

Seed: Gerbera seed is available from numerous suppliers. There are 8,000 linear-shaped seed per ounce.

Sow seed evenly in rows in germination trays using sterile, well-drained, peat-lite media. Cover seed with ⅛ inch vermiculite or media and mist in. Germinate at 68 F soil temperature. During the germination period avoid direct sunlight and maintain uniform soil moisture levels. The seed will germinate in seven to 10 days.

After emergence, treat seedlings with a fungicidal spray such as Benlate to prevent foliar disease. Seedlings will be ready to transplant four to five weeks from sowing. Early flowering cultivars finish in 13 to 14 weeks, later flowering cultivars take 15 to 17 weeks.

Plugs: Seed germinate in seven to 10 days at 68 F soil temperature. Grow plugs at 70 to 75 F. Use supplemental lighting, preferably HID, at 600 to 1,000 footcandles for 14 to 16 hours per day to promote strong root development.

After the first leaf expands, start a constant liquid feeding program using a complete fertilizer, such as 15-16-17 at 75 ppm nitrogen. At the second true leaf stage, increase to 125 ppm nitrogen. If the seedlings are grown in a plug tray with 200 or more cells, reduce fertilizer applications to twice per week. Also, if a slow release fertilizer has been incorporated into the media mix, apply 75 ppm nitrogen throughout the entire production period in the plug tray.

Growing on

Transplanting: Pick seedlings from the germination tray four to five weeks after sowing or when two true leaves have developed. At this stage, transplant into 2½-inch pots using a well-drained, peat-lite medium. Monitor watering the next two weeks—overwatering new transplants can cause root or crown rot. When five true leaves have developed (four weeks after transplanting), pot into 4-, 5-, and 6-inch pots and space on 6-, 7-, and 8-inch centers, respectively.

Lighting: Place plants in full sunlight (4,000 to 5,000 footcandles) year-round. Growers in the South and West may need light shade (25 to 30 percent) in the summer.

Gerberas are not dependent on a specific photoperiod to flower, but some cultivars initiate buds earlier, having a higher flower count and more uniform flowering, when exposed to short-day treatments.

Temperature: Grow gerberas warm, maintaining 70 to 75 F day temperatures. Night temperatures should hold at 60 to 62 F. Lower night temperatures and higher day temperatures delay flowering.

Fertilization: Gerberas are heavy feeders. After transplanting, apply 75 ppm nitrogen of the 15-16-17 formula for two weeks. Increase fertilizer rates to 200 to 250 ppm nitrogen to crop finish.

Insects: Aphids, whiteflies, mealybugs, spider mites, leafminers, thrips, and certain caterpillars can infest gerberas. Oxamyl 10G applications four weeks after transplanting will help control these insects. Because it has a local systemic characteristic, Orthene also controls these insects. Vendex 50WP controls spider mites with good residual properties.

Disease: Soil-borne diseases such as Pythium, Rhizoctonia, and Phytophthora can attack the crown and root system. Foliar diseases of gerberas include Botrytis, Alternaria, and mildew. Botrytis can also attack flowers, causing brown spotting on the petals. Alternating between Benlate and Ornalin applications should control that.

99

Lewis Howe is research assistant, Park Seed, Greenwood, South Carolina.

Godetia

Culture notes

May 1990

by Robert G. Anderson

Satin Flower
Family: Onagraceae
Genus, species: *Clarkia amoena* ssp. *whitneyi*

Satin flower, or godetia, has been a seldom-used garden flower for many years. The genus *Clarkia* (*Hortus III* does not split the genus *Godetia* from *Clarkia*) has over 10 species that are wild flowers on the West Coast from California to British Columbia. Satin flower is grown as a greenhouse cut flower in Japan and is receiving renewed interest as a field grown cut flower in California and Europe with the introduction of hybrids. Techniques to grow satin flower as a pot plant have recently been developed.

Satin flower is a cool season, high light plant that resembles a snapdragon with large flowers and a free-branching plant habit. It produces 25 to 40 3-inch, petunia-like flowers at the tip of the primary stem and four to 10 flowers at the tips of branches. Primary stem flowers open four to seven days before laterals; flowers on all lateral stems open within one to two days of each other.

Flower petals have a distinct satiny texture and unusual patterns of striking colors. Individual flowers last five to seven days; individual inflorescences have flowers open for over three weeks. Cut stems last over two weeks in tap water because lateral flowers continue to open.

Pot plants from seed
- **Seed.** Seed germinates in four to five days at 70 F soil temperatures under intermittent mist. Remove seedlings from bottom heat and mist after seven days. Transplant seedlings to cell flats or pots at 14 days.
- **Media and containers.** Satin flower grows well in many growing media. Four-inch pots are appropriate for single plants. Plants from seed lack the necessary uniformity to have two to four plants in a 6-inch pot.

Godetia

- **Fertilization and watering.** Moderate fertilization at 200 parts per million N and K from standard fertilizers is satisfactory; no unusual nutritional needs or deficiencies have been observed. Satin flower is an excellent candidate for scheduled subirrigation in ebb and flood benches. The plant is vigorous, dries out quickly and wilts easily. Young plants are soft and easily knocked down by handwatering, leading to bent, distorted stems.
- **Temperature.** Use 60 to 62 F at night; keep day temperatures 10 to 15 degrees higher. High temperatures reduce plant size and vigor.
- **Lighting.** The key to production is light. In winter, plants flower 10 to 11 weeks after sowing when grown under 600 to 800 footcandles of high pressure sodium lighting for 18 hours per day. Under winter ambient light, plants require more than 20 weeks of lighting. Long day extension with incandescent lamps hastens flowering slightly but plants are tall, weak and spindly. Controlled short day treatment delays flowering slightly over ambient light conditions.
- **Growth retardants.** B-Nine, Cycocel, Bonzi and Sumagic reduce the normal 24- to 36-inch plant height to 8 to 12 inches for pot production; A-rest is not effective. Use two to three applications at two- to three-week intervals of a B-Nine spray or drench at 3,000 ppm, a Cycocel drench at 4,000 ppm, or a Bonzi drench at 20 to 30 ppm. Cultivar and production light level affect height control. Growth retardant treatments delay flowering two to four weeks but don't affect flower size.
- **Insects.** Whiteflies, aphids, spider mites and thrips can be a problem. All can cause severe infestations quickly, so a standard preventative spray program is necessary. Aphid and whitefly populations build up quickly on the relatively hidden lower leaf surface.
- **Diseases.** No Botrytis or powdery mildew problems have been noted on plants grown near other plants infected with these diseases. Bacterial wilt, possibly seed borne, has killed two to five percent of hand watered plants and has been transferred between plants by hand watering. With subirrigation, we've seen no bacterial problems.
- **Cultivars.** Grace Rose Pink, Grace Shell Pink and Grace Salmon from Sakata Seed Co. have been evaluated for pot plant production as described above. Other cultivars are available from other seed companies that will be used for future evaluations.

Pot production from cuttings

To reduce production time for pot godetia from seed, eliminate problems from the floppy, soft growth of seedlings and increase plant uniformity for larger pot plants, try pot plant production from cuttings. We use terminal cuttings with flower buds present. Preliminary trials show these cuttings can be pinched a week after potting, receive one growth retardant treatment and flower four weeks later when grown under supplemental high intensity discharge light.

Terminal or lateral cuttings 2 inches in length root easily. Treat cuttings with rooting hormone, place in rooting media or direct stick in growing media. Keep at 70 to 75 F media temperature under intermittent mist. First roots are visible in 10 days; transplant cuttings in 14 to 18 days. For pot production, use cuttings with flower buds one-fourth to one-half-inch long.

Grow stock plants for cuttings under cool temperatures and high light conditions. Northern greenhouse production for Valentine's Day requires sowing seed by October 15. Grow stock plants at 62 F night temperature and under 18 hours of HID light at 600 to 800 footcandles. Give stock plants a hard pinch to leave five to eight lateral branches five weeks later, by December 1. Harvest cuttings from lateral branches three to four weeks later when the largest flower buds are one-half inch long. Root and transplant cuttings by

January 5 and place plants under the same supplemental light. Pinch out terminal flower buds by January 12 to allow lateral branches to elongate to a uniform height. It may be necessary to treat lateral branches with one growth retardant treatment when they are 1 to 2 inches long.

Southern-grown cuttings shipped north between Christmas and New Year's could be finished for Valentine's Day, four to five weeks, when grown under supplemental HID lighting. Under winter ambient light, cuttings require eight to 10 weeks to flower.

"

Robert Anderson is professor of extension floriculture in the Department of Horticulture and Landscape Architecture, University of Kentucky, Lexington.

Gomphrena

Culture notes

January 1990

by Meg Williamson

Gomphrena (Globe Amaranth)
Family: Amaranthaceae
Genus, species: *Gomphrena haageana* and *G. globosa*

Gomphrena, or Globe Amaranth, is a multipurpose annual that's gaining well deserved recognition, especially for its use as a dried flower. As a cut flower, it supplies countless bouquets of extremely long lasting blooms. As a bedding plant, it flowers beautifully all summer, providing eye-catching color long after many other annuals are languishing in the heat.

Propagation
Seed: Gomphrena is easily grown from medium-sized seeds (12,000 seeds per ounce). Sow in a good, well drained potting medium and cover lightly. Or use a finer medium and sow in plugs. Bottom water or mist seed or plug flats and germinate at soil temperatures of 68 to 72 F. Germination occurs in seven to 10 days, but may be slightly erratic in the *G. haageana* varieties.

Gomphrena

Growing-on

Transplanting: In two to three weeks from sowing, transplant seedlings into a well drained peat-lite medium. If growing for green sales, transplant into 2½-inch pots or packs and grow on another three to four weeks. For plants that will be sold in flower, transplant directly to 3-inch pots or packs and grow on six to seven more weeks. If finished 4-inch material is desired, transplant first to 2½-inch pots and then to the 4-inch pots three to four weeks later, and grow on another four to five weeks. For cut flower production, transplant from 4-inch pots to 6- to 8-inch pots after three to four weeks, then grow on another four to five weeks. If growing field cuts transplant 2½-inch material after all danger of frost has passed.

Lighting: Gomphrena has a high light requirement, with full sunlight (4,000 to 5,000 footcandles) producing the best flowering and plant quality.

Temperature: These plants thrive under high temperature, even up into the 90s. Growing them on the cooler side—68 to 75 F days, 60 to 65 F nights—however, helps minimize stretching and promotes compact growth.

Fertilization: Once seedlings become established, begin a constant liquid feed program using 20-10-20 at 200 parts per million every watering. Continue this throughout the crop time, increasing the concentration to 250 to 275 ppm if the foliage begins to lose color. For plug grown plants, use a lower rate of 75 to 100 ppm beginning at emergence of true leaves and continuing until transplanting, after which the higher rate can be used.

Watering: Allow soil surface to dry slightly between thorough waterings. If Bonzi is used as a growth regulator, plants tend to consume less water.

Insects: Gomphrena is seldom bothered by insect pests. Thrips occasionally rasp the foliage, but they don't harm the tough, papery flowers. If thrips or any piercing-sucking insects become a problem, use Orthene, Talstar, Avid or Safers soap to control them.

Diseases: Few diseases occur on gomphrena, although a leaf spot fungus, *Cercospora gomphrenae* can infect foliage in the South. Use Benlate to control this disease. Two virus diseases, tomato ringspot and hydrangea ringspot, occur but it's unlikely these would be contracted in the greenhouse.

Growth regulators, pinching, special requirements

The variety Buddy is naturally dwarf and needs no growth regulators, while the *G. haageana* and other *G. globosa* varieties definitely need some type of chemical control, especially if grown to flowering size.

B-Nine has little effect on gomphrenas, even when two applications at 5,000 ppm are made. Bonzi gives excellent results with *G. haageana* varieties, but effects are extreme on *G. globosa* varieties.

Apply Bonzi at 62.5 ppm to *G. haageana* seedlings one to two weeks after transplanting, or when they have developed two to three pairs of true leaves and are the size of a dollar coin. Three to four weeks after application, flower buds will appear. Due to the growth regulator, buds form in the crown without stalks, but once the flowers open a week or two later, they will have developed nice sturdy stalks.

A 15.6 ppm Bonzi treatment may be applied to *G. globosa* varieties at the two to three true leaf stage, but some leaf deformation will occur, and plants may need an extra three to four weeks to grow out of it to some extent. Two alternatives are to sell green with picture labels or to grow in 4-inch pots on 10-inch centers.

Cut flowers: Gomphrena flowers dry easily when hung upside down in a warm, dry, dark place with good ventilation. Allow nine to 12 days under average room conditions (68 to 72 F with 30 to 40 percent humidity). At temperatures of 80 to 85 F with humidities of 10 to 20 percent, drying occurs in five days or less.

Fresh gomphrena flowers have a vase life in excess of two weeks. Side shoot flowers that develop on most stems often wilt in water at room temperature if the group is left intact. Therefore, it's best to cut the shorter, individual flower stems for use as fresh cuts.

Cultivars

All the major seed companies offer gomphrenas. Strawberry Fields *Gomphrena haageana* is the first true red gomphrena; it grows 24 inches tall in the garden and bears 1- by ½-inch strawberry blooms.

The *G. globosa* species is the most familiar gomphrena. It also grows to 24 inches and its ¾- to 1-inch flowers come in white, rose, purple and a mixture. Buddy is a dwarf variety that grows 8 to 10 inches tall and maintains a neat compact form. Its ½- to ¾-inch flowers come in purple, rose and white.

Lavender Lady is a Park Seed exclusive that will be offered in the 1990-91 Park Seed Wholesale Catalog. It's similar to the species in habit and other characteristics, but bears lovely, soft lavender blooms.

99

Meg Williamson is research assistant, Park Seed, Greenwood, South Carolina.

Heliconia

Culture Notes

January 1986

by Teresa Aimone

Heliconia (Lobster claw, Parrot's flower)
Family: Heliconiaceae
Genus, species: *Heliconia psittacorum*

Heliconias are just one more exciting addition to the growing store of new and exotic cut flowers. The interesting bracts form an inflorescence similar to that of bird-of-paradise; heliconia flowers are smaller, though, which makes them more readily incorporated into flower arrangements. Heliconia also feature banana-like foliage, long straight stems, and good vase life. And, with proper cultural techniques, heliconia plants are prolific flower producers. Colors range from solid yellow and orange to pink and red.

Production
Propagation: Though heliconias can be propagated from seed, propagation from rhizome division is the quicker of the two methods. Seed germination can take up to two to three months under warm 80 F conditions. Depending on the cultivar, flowering time from seed germination can be one to two years or more.

Heliconias grow from underground rhizomes. Since rhizomes are stem tissue, shoots will be produced from nodes on the rhizomes. Recommendations suggest planning rhizome divisions containing three shoots at a 12-inch on center spacing in production beds. Production time to flowering depends on soil temperature and time of planting; time to flower can range from three to 18 months.

Planting times: Heliconias are native to tropical areas, and they can be planted outdoors in areas with this type of climate (southern Florida, for example). In these areas, planting times can begin in April and may continue through late fall in unprotected beds. Heliconias can be planted year-round in the greenhouse since flowering is not dependent on photoperiod.

Temperatures: Temperatures below 50 F will damage both flowers and foliage of heliconias (other sources suggest maintaining a minimum of 55 F. With soil temperatures of 80 F and up, flowering times from division plantings can approach three months.

Medium: Recommendations suggest planting heliconias in a medium containing a 1:1:1, sphagnum peat, perlite, and sand mixture. The medium should be well-drained since heliconias like lots of water, but not wet feet.

Fertilization: Fertilizers high in nitrogen will promote rapid growth and flowering. pH should not be alkaline; if it is, or if medium is deficient in micronutrients, these situations can be corrected with foliar nutrient applications. Use a surfactant with the foliar applications since heliconias have a waxy leaf surface.

Light: Light intensity has a dramatic effect on the growth and flower production of heliconias. Research has shown that plants grown in full sun produce four times as many flowers as those grown under 65 percent shade. Plants grown under shade will also be taller, obviously, and stems will be weaker than those on plants grown in full sun.

Harvesting: The terminal inflorescence will be produced after four to six leaves have emerged and developed; this number of leaves depends on the cultivar. And, if all cultural requirements have been met, the cuts can be harvested about eight to nine weeks after the terminal shoot emerges.

Final inflorescences can range from 3 to 8 inches across; peduncles (flower stems) will be slender, but strong.

When harvesting the cuts, remove the entire stalk at soil level. Removing the entire heliconia stalk allows more light to reach newly emerging shoots and allows the final end user greater flexibility to work with the longer stalk lengths.

Additionally, any portion of the stalk left remaining will continue to absorb plant nutrients which might otherwise be used in flower production and growth.

Shipping and handling: Heliconia inflorescences will not open further after harvesting, so efforts to force the flowers won't work. Remove the lower stems and foliage after harvesting and before shipping.

Heliconia flowers like high humidity, so place the flowers in water upon receipt to ensure a moist environment.

Shipping the flowers in moist packing material will also help. Store at 55 to 60 F. With proper care and handling, heliconias can yield a vase life of 14 to 17 days.

"

Teresa Aimone, former editor of GrowerTalks *magazine, is with Sluis and Groot, Fort Wayne, Indiana.*

Hemerocallis, Daylily

Culture notes

July 1989

by Russell Miller

Daylilies
Family: Liliaceae
Genus, species: *Hemerocallis* sp.

Beautiful for a day, daylilies continue to gain in popularity, being more popular now than ever before—especially with landscapers and low-maintenance gardeners. As a ground cover and colorful, yet protective erosion control planting, daylilies are very easy to grow and maintain. They are especially favored for their adaptability to different types of soils, heat tolerance, drought resistance and the wide diversity of flower colors, forms, shapes and plant habits.

Branching, higher bud count, durability and flower size have been improved by breeders, and the season of daylily bloom has been greatly extended. The majority of daylilies bloom in midsummer, although there are cultivars that extend the season, blooming from May to frost.

A significant advance in breeding came with the tetraploids, which have double the number of chromosomes to increase the gene factors, resulting in more flower color combinations and forms, increased plant vigor and other positive results. The usual propagation method is by division. Tissue cultured daylilies are another step forward in daylilies. Varieties that may not divide as easily as others can be economically produced from tissue culture.

Growing daylilies from plugs or divisions

Twyford Plant Laboratories Inc. supplies daylilies as Pedigree-Plugs in their recently introduced Collector's Choice program. Growers pot plugs in 1-gallon containers during September and October for spring sales. Finishing time for daylily plugs in 1-gallon containers is 16 to 18 weeks for dormant (deciduous) varieties or up to 48 to 52 weeks for evergreen varieties.

Many other wholesalers supply heavy, one-to-three fan divisions to retailers and rewholesalers. Usually, retailers pot divisions for fall or spring sales in 4½-inch pots or larger containers. For growing-on in the North (cold climate areas), divisions should be planted at least one month to six weeks before a hard freeze. For growing-on in the South (mild winter areas), daylilies can be planted anytime.

• **Use a well-aerated media,** such as a mix consisting of peat, bark and sand at a 1:1:1 ratio. To prevent crown rot, don't plant daylilies too deeply. The soil level should be no more than 1 inch above the crown. Lightly feed in the spring

Quannah is one of several daylilies in Twyford's Collector's Choice program. An evergreen variety, Quannah may be best suited for mild winter climates.

with 5-10-10 or 5-20-20. Avoid a high nitrogen fertilizer; it tends to make scapes too tall.
• **Uniform watering is key** to growing top-notch daylilies. Keep plants evenly moist but not constantly wet, and never let plants dry out. Consistently watering during the blooming season will increase flower size and substance.
• **Optimum growing temperatures** are 80 to 85 F. Many daylilies are cold hardy and can survive temperatures below freezing, although minimum night temperatures of 60 to 65 F are best for continued growth. Daylilies should be planted in no less than half-day sun; optimum light levels are 6,000 footcandles to full sun.
• **Common disease and insect problems** are easily avoided with proper preventative care. Thrips, mites, Botrytis blight, leaf spot and root rot can become problems with daylilies. Fungicide drenches and rodent bait should be applied before overwintering daylilies.

Understanding hardiness

Outside of California, Florida and southern portions of the Gulf states, growers need to carefully consider the regional hardiness and durability of semi-evergreen and evergreen varieties.

Growers, as well as gardeners and landscapers, don't want to have to worry about hardiness with fall planting, and they don't want to replace thousands of dead or injured plants in the spring. Generally, only dormant cultivars or *proven* evergreens should be grown in cold winter climates.

Many interlocking genetic and cultural factors determine hardiness. There aren't just three daylily groups—evergreen, semi-evergreen or dormant—but hundreds of degrees of hardiness, from evergreen to dormant.

There is no absolute, clear cut relationship between hardiness and of dormancy. Evergreens are not truly evergreen where winters are cold enough to freeze leaves repeatedly. Both evergreens and semi-evergreens vary in hardiness, but there are some varieties in all three groups that can be grown in cold winter climates.

Many evergreen and semi-evergreen varieties have proven very hardy in the North and have performed well over the years. There are others, however, that are not hardy enough to withstand the cold Northern winters. There are also some Southern-bred dormants that are not hardy in the North. It's not enough to be dormant—the cultivar must be *hardy* dormant. Additionally, many cultivars listed as dormant in the South turn out to be semi-evergreen and evergreen in the North.

Evergreens, without mulching in fall, may fare well outdoors in the North if enough snow cover remains all winter. It's the open winters with alternating freezing and thawing that make the difference between certain evergreen or semi-evergreen daylilies surviving winter or turning to mush in spring.

"

Russell Miller is a former staff writer for GrowerTalks *magazine.*

Twyford Plant Laboratories, Santa Paula, California, (800) 327-9988; and Andre Viette Farm & Nursery, Fishersville, Virginia, (703) 943-2315, provided technical information for this article.

Hibiscus

Culture notes

September 1985

Teresa Aimone

Hibiscus (Rose-of-China, China Rose)
Family: Malvaceae
Genus, species: *Hibiscus rosa-sinensis*

Hibiscus production is definitely on the rise. Growers are finding that consumers love the vividly colored exotic flowers, the glossy foliage, and the compact habit. Hibiscus are native to East Asia, and the flower forms may be single or double in shades of red, white, pink, orange, yellow, or bicolors; the foliage can have a serrated or smooth margin and be variegated or solid green. The reproductive parts of the flowers, the stamens and pistils, extend out from the petals and add to the dramatic appearance of the hibiscus flower.

At the Ohio Short Course, Herb van der Ende, Burnaby Lake Greenhouses, Ltd., Surrey, British Columbia, gave a talk on "Hibiscus Production." He outlined the cultural procedures he uses in Surrey, 100 miles north of Seattle, and also included some tips for southern growers. "Though the hibiscus is a tropical plant and is commonly used as an outdoor landscape plant in the South, there is a surprising amount of production in the North," Herb says.

Propagation

Hibiscus can be started from cuttings, seed, layerings, or graftings. For *H. rosa-sinensis*, the most common propagation method is from cuttings. And due to the quicker turnaround time, many growers are even foregoing this step and purchasing 2¼-inch liners or prefinished plants.

Herb maintains stock plants grown in ground beds for his cutting source. Pipes buried 15 inches below the soil have 75 F water running through them and provide bottom heat. During the winter, HID lights are used at Burnaby Lakes to maintain cutting production. And since the big demand for hibiscus is during spring and summer months, it's important to keep cutting production high during the winter. Both eye and tip cuttings have been taken at Burnaby Lake, and Herb says that tip cuttings work the best. "They have thicker shoots, and we end up with less cull liners. The eye cuttings have uneven rooting."

When taking cuttings, Herb says they take 3- to 4-inch cuttings with three to four internodes. During maximum production time, cuttings are taken every week.

The cuttings at Burnaby Lake are planted three per 2-inch peat pot and placed in a propagation area with both mist and bottom heat. Herb says the larger leaves are trimmed back so there is no overlap, resulting in less disease. The propagation beds are peat beds covered with 2-mil poly; the poly helps support the pots, and the poly is regularly changed to reduce the risk of disease.

No rooting hormone is required, and Herb doesn't use any. However, a light dusting of the lowest strength hormone will hasten rooting. Soil temperature is maintained at 75 F. Depending on the time of year, rooting will occur in four to 10 weeks. No HID lights are used in the propagation area.

Growing-on stage **Media:** Herb recommends using a coarse, porous mix that has a pH of 6.0 to 6.5. Though no specific mix was mentioned, other sources suggest a 1:1:1, soil, sand, peat or soil, perlite, peat mix.

Containers: Container size depends on your market preference and the amount of time you are willing to spend with the crop. Given adequate light and spacing, most hibiscus varieties respond well to pinching, so the desired plant form and height should be rather easily attained. Again, it depends on your market and desired crop time.

Temperature: During the production stage, hibiscus like to have a night temperature of 65 to 70 F. Hibiscus are at home in tropical areas, and so they respond favorably to high temperatures.

Lighting: Hibiscus also like high light levels, and Herb suggests a minimum of 1,000 footcandles for good growth. HID lights during the spring months will bring the flowers in faster, but perhaps may not be economical due to higher temperatures and higher light during this lime of the year.

Hibiscus

Watering: Never let hibiscus dry out; bud drop and leaf drop will occur. Excessive drying will also cause soluble salts' levels to rise, resulting in burn. In a mix that is well-drained and well-aerated, hibiscus will take lots of water.

Herb has found that hibiscus respond well to both mat and tube watering. He prefers, however, the tube watering since it gives better aeration, lower humidity levels, and less problems with disease.

Fertilizing: Herb begins feeding as soon as the plants have callused in the propagation area. After one thorough leaching with clear water, he uses a CLF program with an ammonium-based feed. Using a 2-tank fertilizer mix (30-gallon tanks), the solutions are applied at a 1:200 ratio. Here are the two mixes Herb uses:

- **Tank A**
  16 pounds ammonium nitrate
  16 pounds potassium nitrate
  8 pounds ammonium sulphate
  16 ounces iron chelate

- **Tank B**
  12 pounds magnesium sulphate
  8 ounces manganese sulphate
  $1/2$ ounce copper sulphate
  $1/2$ ounce zinc sulphate
  $2^1/2$ ounces Borax
  21 fluid ounces 75 percent food grade phosphoric acid
  96 fluid ounces molybdenum stock solution

"A simpler alternate feed program that could also be used would be 20-20-20 in Tank A and an additional nitrogen source, such as sulphate of ammonia, in Tank B," Herb adds. He did warn, though, that sometimes a premixed fertilizer seems to cause flower burn and leaf curl, so you may want to test accordingly.

Pinching: When given proper spacing and high light, hibiscus will break very well after a pinch. Depending on the desired size of the finished product, give the plants one or two pinches. Herb recommends one pinch for a 5-inch and two for a 6-inch. "For best breaking response," says Herb, "the shoots should be woody before pinching. A hard pinch is preferred over a soft pinch. Make the first pinch about three weeks after planting into the final container."

Growth regulators: Hibiscus are Cycocel-responsive, and the timing of the application is tied closely to the pinching times. Generally, only one Cycocel application is needed during the winter; depending on the variety, plants produced in the summer require two applications. After the first pinch, make the application when new shoots are 1 to $1^1/2$ inches long. If a second application is required and a second pinch has also been done, make the second application about one week after the second pinch. Herb recommends not making an application before the final pinch, except in very high light areas. "Otherwise, the plant might grow right through the Cycocel," he says. "In lower light regions, an application before the final pinch can permanently stunt the plant. If the variety is particularly vigorous and a second application is required but only one pinch has been planned, make the second application about one week after the first. If you want to force the plants into flower during bud initiation, you might consider using HID lights right after the Cycocel application. Herb uses a rate of $1/2$ ounce of Cycocel per one gallon of water.

Spacing: A recommended final spacing for 5- and 6-inch hibiscus is 12-inch by 12-inch.

Timing: For Mother's Day production, hibiscus are a 5- to 6-month crop. Mother's Day is a popular selling date for hibiscus, and they are also an excellent fill-in crop for summer. Generally speaking, without supplemental lighting, hibiscus do not perform or flower well after October 1.

Insects and diseases

Whiteflies, red spider mites, and aphids will all attack hibiscus. Pentac and Plictran can be used-on a preventative basis. Herb cautioned that applications of Lannate will cause leaf spot or leaf loss.

For disease control, recommendations include Captan, Benlate, Daconil, and Terraclor. For all pesticides, follow label recommendations.

Varieties

Hibiscus are enjoying a huge surge in popularity, so the new varieties are flying thick and fast. The wide range of colors and bicolors, singles and doubles, allow the grower to gear toward the market demands and still be certain that needs can be met.

"

Teresa Aimone, former editor of GrowerTalks *magazine, is with Sluis and Groot, Fort Wayne, Indiana.*

Hosta

Culture notes

June 1989

by Russell Miller

Pot-grown hostas
Family: Liliaceae
Genus, species: *Hosta* sp. Frances Williams

Hardiness, diversity, longevity with freedom from trouble, as well as beauty and charm—pot-grown hostas offer the opportunity to add a special uniqueness to retail benches from spring to summer.

Frances Williams hosta is an early blooming hosta with pale-lavender flowers, round and puckered, variegated leaves with a wide, yellow irregular margin and blue-green base. Frances Williams performs best under partial shade in moist soil, but also performs well indoors. Liners, bare roots, tissue-cultured hostas and finished plants are available from commercial growers from spring through fall.

Hostas for spring or fall delivery

When you receive hostas in spring or fall, immediately unpack the boxes and check to see that the soil is slightly moist, and moisten the soil if it isn't.

Hosta

Planted in 1980, under normal weather conditions, we expect this planting of Frances Williams hosta to offer the same amount of enjoyment in the year 2000!

Most hostas shipped in spring have been stored in coolers and appear to have freeze-damaged petioles (leaf stalks), just as garden hostas do after a killing frost. The petioles can be carefully trimmed away so as not to damage the new eyes. Allow any remaining petiole stubs to dry, after which they can be pulled or cut off.

Occasionally surface mold develops on the soil or in the petiole tips. This is not harmful and is caused by the high humidity necessary in cold storage.

Pot hostas in containers sized proportionally to their description. Hostas with large or medium foliage size at maturity can be potted in gallon containers; small to dwarf foliage sizes can be potted in 2-quart containers.

Soilless mixes such as Metro-Mix or Sunshine Mix provide a good medium. Root pruning isn't advised. Containers allowing natural air pruning eliminates detrimental root spiraling.

Potted hosta can be overwintered the same as other potted perennials. After applying a fungicide drench, we offer three overwintering suggestions based on our Midwest climate:

- Place pots in a cold frame and cover with a layer of microfoam and a layer of white copolymer, possibly turning larger pots on their sides. Space rodent bait liberally throughout the area. It's very important to remove the covering at the critical time in early spring. Although low in cost, once the covering is removed in spring, plants are exposed to extreme temperature changes along with wind and moisture. This can cause foliage and root damage and possibly plant loss.
- Unheated overwintering structures covered with a white copolymer are ideal. Place pots inside (with rodent bait spaced liberally throughout the area) and cover with microfoam. The microfoam is removed in early spring, but the white copolymer can remain in the houses for additional time.

The ends of the structures can be opened for ventilation, and as spring weather improves, the white copolymer can be removed and replaced with 50 percent shade cloth for continued growing. A 70 percent shade cloth is recommended for blue hostas as it helps hold the blue color longer into the season.

- Hostas can be overwintered in polyhouses with minimum heat, around 35 F. Watch for insects and slugs. Normally these plants develop faster than in their natural environment. Advanced foliage on hosta can't be hardened off and damage can occur if exposed to low temperatures. We don't recommend a warm house as hostas require a prolonged cold treatment to break dormancy.

Hostas change throughout each season in their four- to six-year maturing process. Intensity of most leaf colors and variegation develops from early spring throughout the growing season. Some plants produce their most distinctive colors in early spring and then the color diminishes as the season progresses. As a hosta matures, the height increases, leaf margins may widen, some may take on a rounder leaf and also a quilted, puckering appearance.

Do not over-feed or over-water hostas. Hostas in a dormant state do not use much water.

Small amounts of slow-release fertilizer can be applied in the fall. Never apply large amounts of nitrogen as this leads to lush, active growth that could result in the loss of plants. Hostas can be fertilized in spring even before dormancy breaks with a top-dressing of slow-release fertilizer or a balanced liquid feed containing minor elements.

Recently divided hostas and hostas from tissue culture may be slightly off their normal cycle, and flowering times can be different than normal. On rare occasions, a flower spike may emerge in spring before any leaves do. If this occurs, break the spike off immediately to encourage new shoot development. If new shoots are rising up through the old dead leaves, the old leaves may need to be carefully removed.

Diseases and insect pests of hostas are minimal and can be easily avoided with proper care.

"

Russell Miller is a former staff writer for GrowerTalks magazine.

Walters Gardens Inc., Zeeland, Michigan, supplied the information for this article, and can be contacted by calling (616) 772-4697.

Hydrangea

Culture notes

December 1984

by Teresa Aimone

French hydrangea, Tea-of-heaven
Family: L. Saxifragaceae Genus, species: *Hydrangea macrophylla*

Hydrangeas were first introduced to England from China in 1789. Native to the Japanese island of Honshu, most of the breeding work has been done in Europe.

Hydrangeas require large amounts of bench space, and must undergo both dormant and forcing periods before flowering. Crop time, from unrooted cutting to finished product, is approximately one year.

Hydrangeas are beautiful potted plants, however. The pink, mauve, or blue inflorescences are common sights in florist shops and garden centers at Easter and Mother's Day. The colorful, showy parts of the inflorescences are the sepals, not the flower petals.

Imitating nature

Hydrangeas in their natural outdoor environment follow this growing sequence: During the spring and summer, flowering occurs. In late summer and early autumn, shortening photoperiods and cooling temperatures cause flower formation in the buds. At this time, all visible growth stops and will not begin until the dormant period passes. Hydrangeas remain dormant through the winter until warming temperatures bring about nature's "forcing" process and subsequent blooming. The cycle repeats.

Growers can either start the crop from unrooted or rooted cuttings and replicate natural conditions in the greenhouse, purchase started plants and see the crop through its dormancy and forcing periods, purchase started plants and force them, or purchase prefinished material.

Dormancy and flower Initiation As stated earlier, flower initiation occurs naturally during the late summer and early fall. Propagators can take advantage of nature's ways and grow plants outdoors during this time for flower initiation. If hydrangeas are grown in the greenhouse, the cool temperatures and short photoperiods may have to be replicated. Maintain a night temperature of 55 to 60 F for six weeks. If extraneous night light is a problem, blackcloth will have to be used-simply follow the natural daylengths. During the day, it is especially important to maintain high light intensities.

Flowers should have initiated at least six weeks prior to the start of the cold storage. Dissection and inspection of the buds with a hand lens will determine if flower formation has occurred. Early flowering varieties may set flower buds as much as four weeks earlier.

Cold storage

Once complete flower initiation has occurred, plants are ready to be placed in cold storage. Cold treatment is necessary to break dormancy; if plants are placed in cold storage too soon, the cold treatment will not affect dormancy, and plants will still be dormant after the treatment. Approximately six weeks or 1,000 hours of cold treatment are required to break the dormant stage. Try eight weeks for early forcing.

For long-term storage, maintain 32 to 35 F; for early forcing and longer cold storage, warmer temperatures are better. 40 to 45 F is considered optimum. Don't go above 50 F.

Applications of gibberellic acid

Gibberellic acid (GA) may substitute for cold treatment. The benefits of doing so are not that great, however, since GA applications may produce excessive elongation on tall varieties, and the amount of forcing time gained is minimal.

If needed, GA can be applied four times (no more) at weekly intervals at a rate of five ppm. These applications are continued (again, no more than four times) until growth resumes. A five ppm solution can be made by mixing one tablespoon 1 percent active ingredient GA-3 in four gallons of water. A wetting agent will help. This treatment should only be used when necessary. Applications should begin as soon as it becomes evident proper growth is not occurring. Another technique used to break dormancy during cold storage is to use mum lighting—either intermittent or continuous. Growth usually begins in less than a week.

Forcing

Plants should be in their final containers at the beginning of forcing. Bring plants into greenhouses set at the proper forcing temperature. 59 F night temperature and 64 to 65 F day temperature are normally utilized. Maintain these temperatures for approximately 80 to 100 days. Depending on cultivar, flower buds should be visible after four to six weeks of forcing. A good rule of thumb to help keep track of the development of the buds (at 60 F) is this: Flower buds should be the size of a pea

Blue Danube hydrangea

after eight weeks; after six weeks, buds should be the size of a nickel, and four weeks before flowering date, the buds should be a little bigger than a half dollar. If the crop is slow, night temperatures can be increased up to 68 to 70 F, but flower buds may be aborted.

Hydrangea

After buds are visible, reduce the night temperature to 56 F. Lower night temperatures will promote the development of long strong stems, good sepal color, and large inflorescences. The temperatures can be adjusted accordingly depending on the stage of inflorescence development. No matter how the crop is produced, finish at cool temperatures.

Finishing time depends on cultivar. Table 1 contains cultivars listed by sepal color and response group or finishing time.

Propagation

A vegetative hydrangea cane can produce many different types of cuttings: terminal cuttings, butterfly cuttings (2-leaf/2-bud cuttings), and 1-leaf/1-bud cuttings. The first two types of cuttings are the best to use. A 1-leaf/1-bud cutting involves splitting the stem in half lengthwise-this provides a good area for disease to enter. Crops produced from this type of cutting require an additional three weeks of production time.

The propagation medium should be on the acid side (a 1:1, peat:sand medium is excellent). Be sure to have good water-holding capacity in the medium-hydrangeas have high water requirements. Maintain 60 to 70 F soil temperature and 55 to 60 F air temperature. Light intensity should be maintained at 2,500 to 3,000 footcandles. Feed with 120 ppm 30-10-10 every week. Continue until cold storage begins.

Rooting should occur in two to three weeks. Either a #1 or #2 rooting compound will accelerate rooting. After rooting occurs, shift to 3-inch pots. Later shifts to larger pot sizes depend on the number of buds or "eyes" on the cuttings. Three-eye cuttings (terminal cuttings) should be finished one cutting per 8-inch container; butterfly cuttings (two leaves/two buds) work best one per 6- or 7-inch container, and 1-eye cuttings (one leaf/one bud) are best one per 5-inch container. This shift to finishing pot size should occur at the beginning of the forcing period. Hydrangeas develop heavy root systems, so the soil mass may have to be loosened up a bit prior to planting to encourage new root growth.

Media and fertility programs

Even though hydrangeas have high water requirements, they do not perform well in heavy soil mixes. A 1:1, peat:perlite mix works well.

Certain hydrangea cultivars can change sepal color from pink to blue by the addition of aluminum sulphate. White cultivars will not appreciably change color no matter what the aluminum content is. Sepal color is determined by the availability of heavy metal ions in the soil, and aluminum is a heavy metal ion. In addition, the availability of aluminum to the plant is determined by the pH of the medium. The pH becomes especially important during the forcing period when hydrangeas are heavily fertilized and the color change would be initiated.

For pink sepals: Maintain a medium pH of 6.0 to 6.5. Feed during the winter forcing period with 25-10-10 (or 30-10-10) at 200 ppm. During summer months, feed rooted cuttings weekly with 120 ppm 30-10-10.

Spurway soil tests should yield the following results for maintaining pink-sepaled varieties: 30 to 50 ppm nitrates; six to eight ppm phosphorous; 15 to 25 ppm potassium, and over 100 ppm calcium. Sepal tissue should contain 200 ppm aluminum and have a pH of 6.0 to 6.2.

For blue sepals: Maintain a medium pH of 5.5. Feed during the winter forcing period with a low phosphate fertilizer such as 25-5-30 at 200 ppm. Summer feeding is the same as for pink cultivars. Apply aluminum sulphate drenches during the last six to eight weeks of production at the rate of $1/2$ ounce per gallon. Repeat every 10 to 14 days.

Cultivar	Sepal color	Response group (total weeks forcing at 59 F night plus last two weeks at 54 F night)
Kuhnert	blue, pink	13
Improved Merveille	pink, red	13
Merritt Supreme	blue, pink	13
Merveille	pink, red	13
Regula	white	12
Rose Supreme	blue, pink	14
Soeur Therese	white	12
Triomphe (Strafford)	pink, red	14

Table 1. Cultivars listed by sepal color and response group. Table from *Introduction to Floriculture*, edited by Roy Larson, North Carolina State University. ©1980 by Academic Press, Inc. Hydrangea chapter by T.C. Weiler, Cornell University.

Spurway soil tests should yield the following for blue sepals: 23 to 30 ppm nitrates; one to three ppm phosphorous; 25 to 45 ppm potassium, and over 100 ppm calcium. Serial tissue should contain 950 ppm aluminum and have the same 5.5 pH as the medium.

The pH of alkaline water can be acidified by using citric, sulphuric, nitric, or phosphoric acid at one pound per 10 gallons during August and early September. After cold storage and during forcing, no more than three applications of acids are recommended. If blue varieties are desired, it's best to avoid using phosphoric acid and try another acid source.

Pinching

If hydrangeas are produced from cuttings, remember it will be a 1-year crop if pinched plants are desired. For cuttings taken March through May of the previous year, pinch one time between June 1 and June 15 for a 5- to 6-inch Easter hydrangea. For an Easter hydrangea started from cuttings taken in June and July, do not pinch. A standard recommendation is not to pinch after July 15 for next year's Easter crop.

Growth regulators

B-Nine sprays during winter forcing help keep plants short. Apply one foliar spray of 2,000 to 5,000 ppm two to four weeks after the start of forcing (remember, plants should be in their final containers). Summer growth is regulated by a 5,000 to 7,000 ppm B-Nine spray. Treatment is made when newly emerging shoots are 1 to 1½ inch long. A second application, at the same rate, should follow two to three weeks later. If growth retardants are applied after early August, the effect will carry over to finished plant forcing, and flowering will be delayed.

Chemical defoliation

Prior to the cold storage period, many growers will defoliate their hydrangeas. This is especially good practice if cold storage space is limited. The lack of foliage helps prevent Botrytis that may occur at tight spacings and cold temperatures. A spray of 1 percent butyne-diol at one pound per 12 gallons water will cause defoliation in about seven days. Chemical defoliation will not break dormancy.

Physiological problems

"Blind" shoots (no flowers on forced shoots) is due to poor fall growing conditions: Low light levels, cool temperatures brought on by premature cold storage, and early defoliation will all lead to blind shoots.

Iron chlorosis is another problem hydrangea growers may have. This problem may be more prevalent in pink varieties since they are grown at pH of 6.0 to 6.5, and iron becomes less available at a higher pH. To correct the problem, apply three applications of iron sulphate during the summer months at three pounds per 100 gallons. Chlorosis is common at the start of forcing—especially if root growth is poor. Check the root system (a practice that should occur on a regular basis) before applying iron.

Insects and diseases

Aphids and red spider mites can be regular pests of hydrangeas. For red spider mites, use Kelthane or sulfur bombs. For aphids, Thiodan, Meta Systox-R, or Malathion will work. Follow label directions for rates. Growers should exercise caution when using some insecticides. Though Malathion will kill aphids on hydrangeas, the chemical can also produce yellow spots on the foliage. Parathion will produce the same damage. Aramite and Lindane can cause chlorosis on some hydrangea varieties.

Botrytis blight *(Botrytis cinerea)* can be treated with protective sprays of Botran 75 percent WP at ½ pound per 100 gallons on a weekly basis. Leaf spots caused by organisms such as *Phyllosticta hydrangeae* and/or *Septoria hydrangeae* can be prevented with sprays of Zineb 75 percent or Daconil 2787 76 percent WP at 1½ pounds per 100 gallons. Spray at 10-day intervals. Powdery mildew *(Erysiphe polygoni)* can be treated with Karathane 25 percent WP sprays at four ounces per gallon every 10 days.

Root rots caused by *Pythium* spp. can be prevented by using sterilized media and avoiding overwatering. Drench established plants with Lesan 35 percent WP at eight ounces per 100 gallons-once. Root rots caused by *Rhizoctonia solani* should be treated with Terrachlor 75 percent WP at four ounces per 100 gallons.

A rosy future

Though hydrangeas have had their high and low points in popularity, they're a crop that is sure to be around for many years to come. Spectacular blooms on well-grown plants make production well worth the effort.

99

Teresa Aimone, former editor of GrowerTalks *magazine, is with Sluis and Groot, Fort Wayne, Indiana.*

Culture Notes

February 1987

by Debbie Hamrick

Polka Dot Plant
Family: Acanthaceae
Genus, species: *Hypoestes phyllostachya*

Polka dot plants are known by a lot of aliases—measles plant, flamingo plant, freckle face, and pink dot—all descriptions for the bright pink coloring on the leaves of this Madagascar native. Also commonly known by its genus name, *Hypoestes*, polka dot plant is guaranteed to spruce up any combination basket or planter. Small pots (4-inch) of the bright, cheery, attention-getting foliage are turning up more frequently in the US for mass market sales. *Hypoestes* is a mainstay in the European pot plant market. It is showing its pink leaves more frequently in US retail outlets as well.

Hypoestes culture is a snap for growers. The plants grow like weeds in Costa Rica, where world-renowned plant breeder Claude Hope first developed *Hypoestes phyllostachya* into a species suitable for pot culture.

Propagation

Cuttings: *Hypoestes* roots readily. Start with cuttings about 3 inches long. Root plants in a soilless media with 30 to 40 percent peat moss. Because cuttings are soft, maintain a high relative humidity with mist. To avoid root rot problems, keep media moist, but not soaking wet. Bottom heat helps to dry media. Cuttings begin to root in six to eight days, and are ready for transplanting in three to four weeks. From direct-stuck cuttings, *Hypoestes* can be a six-week crop in the summer.

Seed: *Hypoestes* seed is readily available from a number of suppliers. Seed is relatively large, 22,000 per ounce. Sow seed in rows, and cover lightly with 1/8 inch media or vermiculite. Use a well-drained peatlite germination media. Seed will germinate in three to four days at 70 F to 75 F. This sowing will minimize transplant loss. Use mist, a germination chamber, or cover flats to maintain humidity. (Uncover flats as soon as seed begins to germinate to prevent stretching.) Seedlings will be ready to transplant in 14 to 21 days from sowing (when the first true leaves appear). From bareroot transplanted seedlings, *Hypoestes* is a nine- to 10-week crop.

Plugs: Seed will germinate in three to four days at 75 F. Double or triple sow for easy transplanting. Grow plugs warm at 70 F to 75 F under supplemental light (600 footcandles for 16 hours a day). Apply Cycocel at the first true leaf stage. Feed with a peatlite soluable fertilizer such as 15-16-17 at 200 ppm or

Hypoestes

less. Plants are ready for transplanting in four weeks, but will hold well in cells. Finishing time for 4-inch (one or two plugs per pot) is four to five weeks in summer.

Planting

For full 4- or 4½-inch pots, use three rooted cuttings or three to four bareroot seedlings per pot. Any well-drained soilless media is fine. For pack sales or for dish gardens, combination pots, or mixed baskets, transplants may go into cell packs for growing on.

Growing on

Light: To really bring out the pink color in leaves, maintain high light levels (up to 2,200 footcandles when plants are established). Growers in the Sunbelt and West may need to shade plants year-round. If leaves start to curl, lightly shade plants. Full light intensity will help in maintaining plant compactness. In northern areas, growers may wish to light seedling crops to promote both compact growth and shorten crop time. Use 600 footcandles for 16 hours daily.

Temperatures: Maintain growing temperatures of 70 F to 75 F for shortest crop time and best leaf coloring.

Fertilizer: *Hypoestes* is not a heavy feeder. A 175 to 200 ppm 15-16-17 soluble formulation or a 200 ppm 20-20-20 soluble formulation once a week is all they need.

Growth regulators

Hypoestes is a vigorous grower. Plants respond well to Cycocel. This regulator also helps to deepen leaf color. Recommendations are that Cycocel may be applied twice at 750 to 1,000 ppm. The first application should be made once plants have established themselves after transplanting (when seedlings are 2 to 2½ inches in diameter). The second application may be made two weeks later. *Hypoestes* also responds well to B-Nine.

Growth regulators are essential to maintain plant height. They will also add flexibility to the shipping schedule. *Hypoestes* is not a plant that can sit on the bench when it's ready to ship. Growth regulators ease this situation somewhat by making a two-week order pulling period.

Varieties

Betina: This European cutting variety features large leaves measuring about 1 inch wide by 2 inches long. Leaves are dark purple red. Plants are spreading, rather than basal branching. One cutting is sufficient for a 3½- to 4-inch pot.

Pink Splash: This seed variety grows 8 to 10 inches tall. It's the most dwarf variety to date from seed. Plants are basal-branching and leaves are heavily mottled with bright pink. Pink Splash ships well over moderate distances. (We hear a new selection that's even more colorful and more compact is in the works for release later this year.)

Polka Dot Plant: Mound-shaped plants grow to 12 inches. Leaves are sprinkled with pink dots.

White Polka Dots: Plants grow to 12 inches. Dark green foliage is distinctly marked and sprinkled with white polka dots. Plants are mound-shaped.

"

Debbie Hamrick is publisher/editor of FloraCulture International *magazine and international editor of* GrowerTalks *magazine.*

Impatiens, New Guinea

Culture notes

October 1989

by Edward P. Mikkelsen

New Guinea impatiens
Family: Balsaminaceae
Genus, species: *Impatiens* x *hybrida*

New Guinea Impatiens weren't known to the gardening public before 1972. Since their introduction, sales have climbed at an annual rate of over 15 percent. New Guinea impatien sales increased about 15 percent this year over 1988. Performance for the consumer is unmatched by any other flowering product.

Equally important are the many advances made by breeders. Plant habit and self-branching are improved. Many flower colors are available in combination with assorted leaf types. Characteristics, such as bicolored flowers, continue to add variety.

Propagation: seed and vegetative

Sow seed late January to early February for sales in late April/early May. Sow seed thinly (five per inch) in a well-aerated mix. Maintain 70 to 75 F soil temperatures. Cover seed lightly with 1/8 to 1/4 inch media and water thoroughly.

Without mist, germination is 10 to 14 days. Under ideal conditions in a germination chamber, germination is seven days. Transplant seedlings into cell packs for establishing on the 28th day or after first true leaves appear.

Vegetative propagation is restricted to licensed propagators. Most growers buy in rooted cuttings. Begin with quality cuttings that are disease and insect free. The best do not have flower buds. If cuttings are budded with no vegetative nodes below the tips, the plants grow slowly and branch poorly. If cuttings are budded but have vegetative nodes below, then plants should perform satisfactorily.

General culture

Containers: Match varieties to container size. Don't use a short variety such as Dawn in a hanging basket, especially if you are growing cool. Likewise, don't grow a vigorous variety such as Enterprise in a 4-inch pot.

New Guinea impatiens are ideal for a wide variety of container sizes: 4-, 5-, and 6-inch, hanging baskets and patio planters and urns. Use one rooted

Impatiens

cutting per 4- or 5-inch pot, one to three cuttings for 6-inch pots, three cuttings for an 8-inch basket, and four to five cuttings for 10-inch baskets.

Prevent over-watering of small plants in large containers by establishing plants first in 4-inch pots and transplanting into larger containers.

Media: New Guinea impatiens have been successfully grown in a wide range of media—soil-based, straight peat or peatlite mixes. Use disease-free media with good aeration.

Temperatures: After transplanting, keep night temperatures 65 to 68 F until plants are well established. At this point lower night temperatures to reduce stem elongation.

Most newer varieties don't require this and temperatures below 62 F slow the crop without benefiting plant habit. Flower initiation occurs best between 65 and 75 F.

Fertilizer: Plants require moderate to heavy fertilization with a minimum of 250 ppm N, P, K at every watering. They require low concentrations of minor nutrients. If the medium has minor nutrients, do not use them in your fertilizer or vice versa.

Minor nutrient toxicity has occurred when used both in soil and in feed. Minor nutrient toxicity appears as brown or black spots on leaves. Severe cases show die-back starting at the plant tip and moving down the stem. Stems have dark areas that are firm in contrast to being soft as with rot. Plants exhibiting these symptoms should be tested for high levels of minor elements and virus.

Crop time: Depending on season and locality, crop time varies. Allow seven to 10 weeks for 4-inch pots; 10 to 14 weeks for 10-inch hanging baskets.

Pinching: With the newer, self-branching varieties, a pinch isn't needed. Properly space pots to prevent stretch.

Diseases and insects

Control disease with good environmental practices. Daconil and Chipco are effective for Botrytis control. Subdue and Truban are effective against Pythium and Phytophthora.

Impatiens

Two-spotted spider mites and thrips are the most important insect pests. A vigorous thrips control program is a must if this insect is present, since New Guinea impatiens can become infected with TSWV. Pentac controls two-spotted mites. For thrips, use Avid, Vydate, Lannate or Thiodan. Chemicals known to be phytotoxic to New Guinea impatiens include Temik, Kelthane and Karathane.

Breeders are providing plant material free of Tomato Spotted Wilt Virus. Plants become infected by thrips infestation at some of the secondary propagators and/or growers. Thrips control programs initiated by propagators and growers have already reduced the incidence of TSWV.

Varieties

Sunshine series: Bred by Mikkelsens, Ashtabula, Ohio. A popular line. Series features widest variety of colors and plant types available.

Kientzler series: Bred in Germany and from Paul Ecke Poinsettias, Encinitas, California. New in the past two years. Available in wide color range.

Bull series: Bred in Germany and from Fischer Geraniums, U.S.A. New this year. Plants more vigorous, well-suited to large container and basket production. Seven colors available.

Celebration series: Bred in Costa Rica and from Ball Seed Co., West Chicago, Illinois. New this year. For 4- and 6-inch pots. Seven colors available.

Tango: First variety from seed. Orange flowers. Plants grow 24 to 30 inches in the garden. Suited for 4-inch pots or larger containers. An All-America Selections Award winner.

"

Ed Mikkelsen is president and director of Research and Development, Mikkelsens Inc., Ashtabula, Ohio.

Impatiens wallerana

Culture notes

November 1990

by Teresa Aimone

Family: Impatiens Balsaminacese
Genus, species: *Impatiens wallerana*

Impatiens have blasted out their niche as the top selling bedding plant. While other classes may have had a hit-and-miss shot at the first place slot, a noted industry survey states that impatiens have held the lead as the most increased bedding plant for the last 20 years. Impatiens hold the top spot in

Impatiens

Left: Blitz Light Pink impatiens. Above: Impulse Mix.

popularity for several reasons: increased turnover, color selection and easy adaptability to climates and a variety of containers.

Culture

- **Germination**: Seed count varies from 35,000 to 60,000 seeds per ounce. Differences vary per color; check with your local salesman to determine required production quantities.
- **Temperature**: Optimum media temperature is 72 to 75 F. Temperatures of 78 F and higher can create a thermal dormancy (seed won't germinate until lower temperatures are provided), or seed can die as a result of these higher temperatures. Provide efficient cooling and misting systems in germination areas that maintain a constant, optimal temperature.

Monitor temperature throughout the germination area, not just in the soil medium, to be certain desired uniformity is being achieved. Temperatures can vary significantly just a few inches away from the plug flat.

- **Media**: Keep the mix as coarse and well-aerated as possible; 70 percent coarse peat and 30 percent perlite is recommended. pH should be 5.8. Avoid saturating the medium. This cuts down on oxygen getting to seed and causes an ethylene build-up.
- **Soluble salts**: Keep salts levels as low as possible. Shoot for a reading of 1.0 $\mu mho/cm^3$ from a 2:1 soil solution.
- **Light**: Impatiens are light-dependent for germination. Don't bury seed in medium; this limits oxygen and is especially detrimental if you're keeping the medium very wet. If you wish to cover seed, provide a very thin covering that will keep moisture on seed and still allow light in. If you leave seed uncovered, keep a thin film of moisture on seed during germination.
- **Air circulation**: Germination areas can be perfect breeding grounds for diseases. Good air circulation is vital, not only for disease prevention, but to

Impatiens

keep temperature evenly distributed.
- **Moving to Stage 2**: Obviously, not all series germinate at the same rate. Remove germinating plug trays from Stage 1 conditions as seedlings emerge and keep notes and records for your future reference. Time differences you note may only be in hours.

Growing-on

Transplant plugs as seedlings begin to crowd or stretch. If you're using a 406 plug tray, this usually happens about 5 to 7 weeks after sowing.
- **Media**: Use a light, well-drained medium, similar to the one used for germination. Use a finer grade of peat, as this will hold moisture better and be particularly beneficial to increase shelf life at retail. Keep pH 6.0 to 6.2.
- **Temperature**: Maintain 64 to 68 F until plugs are established. For Stage 4, the selling or holding stage, drop temperatures to 60 to 62 F.
- **Fertilization**: For the first 10 days after leaving the germination area, feed weekly with a balanced N:P:K feed at 50 parts per million. After this, increase to 100 to 150 ppm every second or third watering for about 2 weeks, until Stage 4. Plants should have a well-proportioned height and foliage canopy in relation to their container and be well-budded or just starting to flower. Keeping plants too wet and too lush will seriously slow flowering and reduce shelf life. Reduce water and feed at this stage to push plants into flower and harden them off for retail sales.
- **Crop time**: Pack production normally takes 10 weeks; add 2 weeks for baskets or other large containers.
- **Insects and diseases**: Impatiens have been called the next Typhoid Mary of Tomato Spotted Wilt Virus because they often predominate production space during the bedding season. Thrips are insect vectors of TSWV that transmit the disease to impatiens.

So, with a greenhouse full of impatiens, thrips and the disease can quickly spread. There is no cure for TSWV other than destroying infected plants, but you can use a spray program to control thrips.

Do your best to monitor thrips populations closely, and use a preventative spray program. There are several chemicals on the market for thrips control (Dursban, Lannate, Thiodan, Avid); contact your local chemical supplier to start your program.

TSWV appears as concentric ring spots on impatiens. Outer ring edges are black, progressing to lighter brown areas inside. If you suspect infection, isolate plants and have them analyzed, maintaining your spray program in the meantime. Destroy all plants found to be infected.

What's next in impatiens?

Impatiens have been subject to many changes since the first non-hybrid series, the Babies, were introduced in the mid-1950s. What's next is up to growers and their customers who guide breeding companies in the proper directions. Voice your opinion in what you'd like to see happen in North America's most popular bedding plant.

99

Teresa Aimone, formerly editor of GrowerTalks magazine, is with Sluis & Groot, Fort Wayne, Indiana. The author would like to thank David Koranski for his help in preparing this article.

Kalanchoe

Culture Notes

October 1986

by Teresa Aimone

Kalanchoe
Family: Crassulaceae
Genus, species: *Kalanchoe blossfeldiana*

Kalanchoes were introduced in Germany in the early 1930s. Since that time, much breeding and development have been done on this crop to make them more attractive to the consumer and more adaptable to growing in the US. In this country, kalanchoes enjoy their greatest popularity in the Southwest.

Propagation
Kalanchoes were once commonly produced from seed. Though they still may be produced this way, the most common method of propagation now is by vegetative cuttings. A major drawback to propagation from seed is time—it takes at least seven weeks before seedlings can be transplanted.

Vegetative cuttings can be produced from stock plants or purchased from a specialist propagator. Container size for stock plants varies with grower preference. Some use 6-inch pots, while others grow stock plants in 1-gallon containers. Provide plants with an initial spacing of 12 inches on center. To ensure that stock plants remain vegetative, provide constant long days. Research has shown that an interrupted night lighting schedule is more effective in keeping plants vegetative than continuous lighting is.

Maintain night temperatures of 65 F. Night temperatures should not exceed 76 F. Stock plants should never be exposed to water stress, as this will reduce cutting production. Plants may be fertilized at every watering using a complete fertilizer such as 20-20-20.

The amount of cutting production will vary depending on cultivar and production techniques. Recommendations suggest renewing the stock two to three times a year to avoid the possibility of premature bud set.

Kalanchoe cuttings are fairly easy to root, so a rooting hormone is not necessarily needed. Cuttings should be 2 to 3 inches in length. A satisfactory rooting medium would be a 2:1 ratio of peat and perlite. Cuttings should begin to callus in seven days, and an adequate root system should be formed within three weeks. Maintain a soil temperature of 70 F. Rooting time and cutting quality is enhanced with the use of intermittent mist.

Purchasing liners: Many growers simply forego the work of producing their own cuttings and purchase rooted 2¼-inch liners from specialist

Kalanchoe

propagators. If liners are used, unpack them immediately upon arrival, and pot them up. Plant the liners fairly shallow; the soil level on the cutting should be the same as the pot soil level or slightly higher. This practice and the following cultural procedures apply to both the grower's own rooted cuttings and purchased liners.

Medium: The medium used should be well-aerated and well-drained—kalanchoes don't like to have the roots constantly wet. A wide variety of media are recommended—1:1:1, peat, perlite, and soil; 5:4:4, soil, peat, and calcined clay; or a 3:1:1, milled pine bark, sand, and peat.

pH: Maintain a soil pH of between 6.0 to 7.0. A pH of 6.5 to 6.8 is best.

Fertilization: Provide kalanchoes with a constant liquid feed consisting of 250 ppm nitrogen, 50 ppm potassium, and 150 ppm phosphorous. Calcium is an important nutrient in kalanchoe production, so either use calcium nitrate as the nitrogen source, or amend the soil in some other way with calcium. Reduce fertilization rates one week prior to short days and two weeks after short days start. Fertilizer concentrations should also be reduced several weeks prior to flowering. Growers may want to reduce their fertilizer concentrations by alternating clear water with fertilizer.

Temperature: Recommended night temperatures are 63 to 65 F. When plants are blackclothed during the summer, don't let the temperature under the blackcloth exceed 78 F, or heat delay problems may occur. If possible, provide good ventilation and air circulation under the blackcloth.

Watering: Keep the medium on the dry side, since kalanchoes are succulent plants and don't like lots of water. Kalanchoes are well-suited to automatic watering—their large leaves makes getting water to the media hard with overhead irrigation. Capillary mats can be used to produce a successful crop, also.

Pinching: Soft pinches are important to produce a well-branched plant. Take a very soft pinch of tips and side shoots. A harder pinch may be required during periods of excessive growth. The pinch can be done from one week before the start of short days to five days after the start of short days, depending on pot size and time of year.

Daylength control

Kalanchoes need 13 hours of darkness per night to initiate buds. Provide long days from September 1 to April 1. Additional weeks of long days may be needed if vegetative growth is not adequate. Recommendations say that adequate long days are provided if the lower foliage touches the container rim just prior to short-day treatment.

It is very important that the short-day treatment is continuous. If one day of treatment is missed out of the week, this will negate the effect of the previous six days of short-day treatment. Visible terminal buds should be detected after five or six weeks of short-day treatment.

Response time varies depending on cultivar; the period of time from the start of short days to flowering varies from nine to 14 weeks.

"

Teresa Aimone, former editor of GrowerTalks *magazine, is with Sluis and Groot, Fort Wayne, Indiana.*

Lilies, Easter

Working the bugs out

October 1989

by Mark E. Ascerno

"Kelthane soak in combination with a monthly fungicide program remains the best choice for beating bulb mites."

Easter lilies are commonly affected by root rot diseases. While growers have consistently used fungicides for their control, Frank Pfleger, Department of Plant Pathology, University of Minnesota, and I have found that as the number of bulb mites increases beyond about 300 mites per bulb, the ability of fungicides to control root rot is severely reduced.

Mite numbers vary in Easter lily bulbs

Mites are commonly present on bulbs shipped from the Pacific Northwest. Their numbers vary from year to year, source to source and bulb to bulb,

Lilies

making it impossible to forecast the need for control in a given year. For this reason, many growers conduct routine "insurance" bulb mite treatments.

The bulb mite, *Rhizoglyphus robini*, is approximately $1/32$-inch in length, with short, stubby legs and a pearly white body. They normally feed on damaged or decaying bulb scales. Their relatively large size makes them easy to spot.

Rhizoglyphus mites have been effectively controlled by soaking bulbs in $1 1/3$ pounds of 35 WP Kelthane (dicofol) for 30 minutes prior to potting. Kelthane also provides a moderate level of root rot control. In our studies, we consistently produced the highest quality Easter lilies when we combined a Kelthane bulb soak with monthly fungicide drenches.

Kelthane's questionable status prompted search for alternatives

The questionable status and availability of Kelthane a few years ago created widespread requests from growers for other materials that might control bulb mites. Unfortunately, with the exception of Temik, alternative bulb mite control materials have not been recently tested. This prompted Frank and me to reopen our studies to look for alternatives to Kelthane.

In addition, due to concerns about chemical disposal and ground water contamination, we made comparisons between materials applied as a pre-potting bulb **soak** versus a soil **drench**. Our feeling was that drenching would avoid the need to discard gallons of spent solution used for soaking bulbs.

Experimental design

Our first trial, conducted during the 1987/88 forcing season, tested Avid 0.15EC, Pentac 50WP, Talstar 10WP and Vendex 50WP on control-temperature-forced Nellie White Easter lilies. We used Avid, Vendex and the current formulation of Kelthane 35WP during a second trial in the 1988/89 season. We used miticides at labeled rates as either a 30-minute bulb **soak** or as a **drench** applied immediately after potting. A 30-minute water soak and a water drench served as check treatments. We applied Benlate/Subdue to selected treatments in both trials.

In 1987/88, we applied Benlate/Subdue monthly. In 1988/89, we applied Benlate/Subdue as a drench immediately after potting with only Benlate being applied monthly. At the end of the experiments data was taken on: number of mites per bulb, root rot severity, dry weight, number of flowers per plant and plant height.

Benlate, Subdue, Talstar, Pentac—Which beats mites?

Although mite numbers were low in 1987/88, some trends were apparent. Talstar and Pentac didn't perform well as either a soak or a drench. Therefore, we decided not to include them in the 1988/89 trials.

We achieved good root rot control with the Benlate/Subdue applications, verifying earlier observations that root rot pathogens can be adequately controlled in the absence of substantial mite numbers. Yellowing of the tips occurred in all treatments receiving monthly applications of Benlate/Subdue. It has been subsequently shown that the yellowing was due to multiple applications of Subdue. Subdue should only be applied to Easter lilies once at planting at the 1-ounce rate. Monthly applications of Benlate, however, did not cause a problem.

Kelthane soak is treatment of choice

Mite numbers were also very low in 1988/89, preventing conclusions about mite control. However, root rot, dry weight, plant height and flower data did provide indications of which material/application combinations might make the best choices, assuming good mite control.

The Kelthane soak produced the tallest, heaviest plants with the least root rot and very good flower development. It was the best overall treatment and since we know that it controls bulb mites, the Kelthane soak remains the treatment of choice.

Other treatments that did reasonably well, in descending order, include Avid drench, Avid soak, Vendex drench and Kelthane drench. The Vendex soak was the most unsatisfactory treatment in nearly all categories.

Due to the low mite numbers in the two studies, it's not possible to suggest a substitute for Kelthane that is known to control mites. Based on the results from 1987/88 and 1988/89, Avid as a drench or soak and Vendex drench appear to be the most promising substitutes if Kelthane is not available. Until such time when the efficacy of these materials on *Rhizoglyphus* mites can be established, Kelthane soak in combination with a monthly fungicide program remains the best choice.

"

Mark E. Ascerno is professor and extension entomologist at the University of Minnesota, St. Paul.

Graphical tracking timetable

October 1988

by Debbie Hamrick

Easter lily height control doesn't have to be a function of the weather or growth retardants. Michigan State's Royal Heins introduced graphical tracking in combination with reversing day/night temperatures. Using this technique on 1988 lily crops, many growers achieved precision height control by manipulating temperature. It works.

Following is a sketch of what graphical tracking and warm night/cool day temperatures are all about. For more information, refer to *GrowerTalks*: November 1987, "The basics on Easter lilies: light and temperature," page 84; December 1987, "Tracking Easter lily height with graphs," page 64; and January 1988, "Precise control of lily height," page 92.

Plotting on the graph

At emergence, graph height of the lily is equal to the height of the pot. From visible bud to flowering, most lilies in commercial conditions double their height.

On the graph set your minimum and maximum allowable finished plant heights for your buyers. Backtrack to visible bud to determine allowable heights at that point in the crop. Connect all the lines.

Lilies

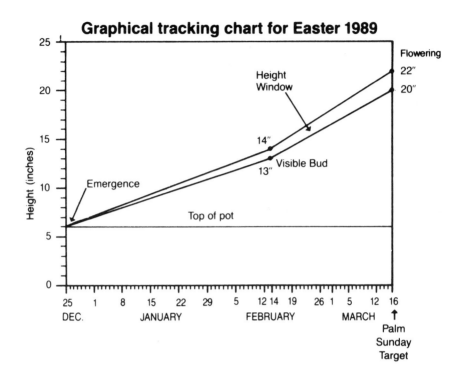

You now have a window of acceptable crop height for any point in the life of the lily.

"The basic concept is the ability to plot your height versus where you should be at any particular point and then do something about height early enough in development so you're not trying to squeeze all your height increases or decreases in the last few weeks," Royal explains.

Prior to visible bud

All leaves must be unfolded before visible bud. Leaf unfolding is a function of average daily temperature. Grade 8/9 bulbs unfold around .09 leaves per day for a 2 F increase in average daily temperature. Larger bulbs unfold slightly faster and smaller bulbs slightly slower.

Determine what leaf unfolding rate you need to acheive your desired visible bud date. For example, on January 17 perhaps 30 leaves are unfolded and your target visible bud date for 1989 is February 14. Refer to the lily leaf unfolding rate graphs to determine what your average daily temperature must be to achieve your leaf unfolding rate. If you determine you need to unfold 1.9 leaves per day on an 8/9 bulb, you'll need an average temperature of 70 F. For smaller bulbs, you'll have to run houses somewhat warmer.

Using temperature to control height

The time it takes for lilies to flower is a function of average daily temperature. Final plant height is determined by how temperatures are delivered. Running cool days and warm nights makes a shorter crop compared to plants grown under warm days and cool nights. This is true even if the average daily temperature for these two regimes is the same.

Lilies grown with 70 F days and 60 F nights will flower at the same time as lilies grown at 60 F days and 70 F nights, assuming the length of the day and night are the same, i.e. 12 hours. But, the plants grown with warm days and cool nights will be taller than the plants grown with cool days and warm nights.

Using this insight on how plants respond to the way temperature is delivered combined with tracking plant height every three to five days on the graph and it's easy to control final plant height.

When plants creep out of the upward acceptable limit, simply run cool day and warm night temperatures. To push plant height upward, run warm days and cool nights. Plants respond almost immediately to changes in temperature delivery, Royal says. "Growers can speed up or slow down plant height on almost a daily basis."

There is a drawback to using cool days and warm nights to control lily height. At a high difference between cool day and warm night temperatures, leaves droop. When the difference is small, or temperatures during the day and night are the same, drooping is insignificant.

Can't achieve cool days? Try pulsing temperatures

Growers who experimented with the cool day/warm night technique last year added a twist to Royal's program. They found that they were most successful when they had tight control in the transition from light to darkness and darkness to light and when they were able to quickly raise or lower temperatures by opening or closing energy curtains.

Could the grower be able to pulse cool day temperatures for a couple of hours in the morning and then raise regular day temperatures but still achieve height control?

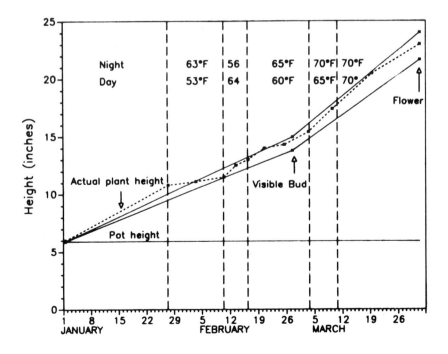

In an experiment to find out, Royal discovered that exposing plants to cool temperatures for just two hours in the morning did significantly decrease plant height.

"We could increase our rate of flower development and still get some height control," he explains. 'I think there's application in some years for it and 1989 might be one of those years."

"

Debbie Hamrick is publisher/editor of FloraCulture International *magazine and international editor of* GrowerTalks *magazine.*

Lilies, other

Culture Notes

December 1986

by Teresa Aimone

Hybrid lilies
Family: Liliaceae
Genus, species: *Lilium speciosum*

Hybrid lilies have risen in popularity over the past years to become a very profitable crop for the grower and an attractive pot plant and cut flower for the end consumer. And part of the attractiveness does lie in their versatility: They can be used for pots or cuts and they can be produced on a weekly basis or for specific target dates.

Hybrid lilies have a relatively short forcing time, and they are a cool-grown plant, so the grower can save money on fuel costs, too. Colorful and unique, hybrid lilies offer growers another profitable item to add to their production lists.

Precooling
Bulbs will be properly precooled when they arrive at the greenhouse. If they cannot be forced immediately, it won't hurt the bulbs if they're precooled longer. Store the bulbs at 34 to 40 F. If bulbs are being forced on a year-round basis, store them at 28 F beginning February 1.

Planting
Lily bulbs can be planted and forced in a wide variety of containers: pots, flats, benches, or ground beds. To allow for adequate rooting, cover the bulbs with a minimum of 2 inches of medium. This will provide enough medium for stem rooting; stem roots allow for absorption of food and water.

Recommendations suggest using 4- to 9-inch bulb size. Bulbs over 9 inches aren't satisfactory. Use the 4- or 7-inch sizes for pot culture, and the 4- to 9-inch size for cut flowers. The bigger the bulb, the higher the bud count.

Medium

Hybrid lilies will tolerate a wide variety of soils and growing media. Whatever the chosen medium may be, it should be high in organic content, very loose, and porous. If the medium is soil-based, it should be sterilized prior to planting to ensure freedom from pests and diseases.

pH

Maintain a pH of 6.0 to 6.5. Low pH has been associated with leaf scorch on lilies, so try to maintain the pH within the recommended parameters.

Fertilization

The bulbs themselves contain lots of stored food, and they should be able to support the plant until the shoot emerges. Recommendations suggest withholding fertilizer until the shoot is 4 to 6 inches tall. Some growers incorporate a 10-6-4 slow release fertilizer in the medium prior to planting. Suggested rate is eight pounds per cubic yard.

Use a liquid fertilizer such as 16-4-12 at the rate of 200 ppm to help deepen foliage color. If a preplant fertilizer has been added to the soil, then just use the liquid fertilizer once every two weeks.

Watering

Water sparingly during the first few weeks after planting. Stem roots are being established at this time, so it's important not to overwater the medium. Maintain uniform moisture content once these roots are well-formed.

Forcing temperature

Maintain a minimum night temperature of 50 to 55 F and a maximum day temperature of 65 F. Hybrid lilies do not require high temperatures, and recommendations suggest that day temperatures not exceed 85 F. Temperatures can be adjusted, depending on how the crop is progressing. If the plants are early, the night temperature can be dropped to 40 F. To speed the crop, increase the night temperature to 60 to 65 F.

Scheduling

The number of days required to force hybrid lilies decreases as the season progresses. Varieties also differ on the number of days to force. Individual growing techniques also affect forcing times.

As a general rule, bulbs planted on December 15 and grown with optimum conditions and 55 F nights and 70 F days will force in 82 days; bulbs planted on January 15 will take 77 days, and bulbs planted on February 15 will take 72 days. These times are approximate; they vary depending on individual conditions.

Physiological disorders

Bud loss: Bud loss can be caused by low light, especially when forced during winter in the north. Some growers supplement natural light with HID lights. Bud loss can also occur during the summer when greenhouse night temperatures often go above 65 F. Research has shown that this problem seems to occur during shoot development, so if bulbs can be placed in an environment where the temperatures can be more closely maintained during this critical period, this may help stop bud loss.

Tip burn: This usually occurs during the winter when cloudy days can be followed by periods of very bright weather. Shade cloth, which helps even out light levels, can help prevent tip burn.

Leaf scorch: Leaf scorch seems to be caused by pH and fluoride levels. Fluoride is found In superphosphate and perlite. Therefore, many growers choose not to use perlite in their mix. Recommended lily fertilizers also are low in phosphorous.

Insects and diseases

Maintaining proper greenhouse hygiene will help control many insect and disease problems. Aphids can often be a pest on lilies. They can be controlled with registered chemicals. Benlate can be used to control Botrytis.

”

Teresa Aimone, former editor of GrowerTalks *magazine, is with Sluis and Groot, Fort Wayne, Indiana.*

Limonium, Statice

Culture notes

April 1991

by Fred Meyer

Statice—Misty Series
Family: Plumbaginaceae
Genus, species: *Limonium latifolium* x *Limonium bellidifolium*

Limonium Misty Series (statice) are hybrids produced by inter-specific hybridization of two limonium species. Produced through plant tissue culture, this successful combination retains the best attributes of traditional, perennial statice. The Misty Series inherits full, branching structure, clean stems without rudimentary leaves and absence of foul scent from *L. latifolium* and height and hardiness from *L. bellidifolium*.

Propagation
This series is supplied by Dai-Ichi Co. Ltd. in the form of stage III plants from tissue culture to ensure genetic stability and uniformity. When Stage III plants arrive, gently empty entire contents of the containers in lukewarm water to soften the agar in which the roots are growing. Keep them in the water until you can easily separate plants without damaging roots.

Put individual plants in 2- by 3-inch pots filled with a media of equal parts washed sand, peat and perlite. After potting, put on two coats of an anti-transpirant, allowing surfaces to dry fully between coats.
- **Acclimatizing:** Place plants in greenhouse with 75 percent shade. Keep moist but not wet; mist lightly several times a day. After about 2½ weeks, expose to 20 percent shade and fertilize lightly. Keep air temperature about 65 to 75 F.

Growing on
- **Transplant:** Transplant into the ground about six weeks after potting Stage III plants. Plant in a 24-inch-wide by 4-inch-high bed, with 16 inches between rows and between plants. Well-drained soil with a 6.5 to 7 pH stimulates root development.

Japanese growers use a reflective poly-cover as a plant mulch to increase productivity, promote vigor and prolong flowering. This series likes some form of mulch to keep soil moist and weeds down, because most of the root system is in the upper soil. Use one or two levels of support on most cultivars, with the

Limonium

possible exception of Misty Blue No. 1, which doesn't need support if grown outdoors.
- **Temperature:** These cultivars have high tolerance for heat and cold, but non-heated or non-cooled greenhouse cultivation is necessary everywhere except California.
- **Fertilization:** Feed 100 parts per million nitrogen of any well-balanced fertilizer. If plants seem a little yellowish, add extra iron in the form of Sequestrene Fe 138. This statice series tolerates high soluble salt levels (EC), two to three times that tolerated by carnations. Plants don't seem to be damaged by EC levels of 2.8.
- **Water:** Keep the top 7 inches of soil moist, not wet, during the growth cycle. Drip tape is preferable to prevent airborne fungus.
- **Harvesting and shipping:** Time from planting to flowering varies from four to seven months, depending on culture and time of planting. Use the liquid solution containing silver thiosulfate and several other chemicals developed by Dai-ichi Seed Co. as a postharvest pulse treatment for freshly cut flowers to encourage continuous opening after cutting. Store at 45 to 50 F. The stems, about 40 inches tall, are sold in bunches of two or three in the United States and singly in the rest of the world.

Varieties

The Misty Series of *limonium* is available in blue, white and pink. Different selections within a color are distinguished by production, flower size and length of flowering time. Currently, Misty Blue No. 1 is the most widely available,

Limonium

with producers in Asia, New Zealand, Australia and Europe under license of Dai-ichi Seed. This product has been produced in the United States for about five years in very limited amounts. Initial response to the hybrid has been very favorable.

Limonium Misty Series is available from Dai-ichi Seed Co. Ltd., Tokyo, Japan [phone 01181-33-427-7173; FAX 01181-33-427-7165].

"

Fred Meyer is president of L.R. Meyer Co., Escondido, California and researcher at University of California, Irvine.

Culture Notes

June 1988

by Jim Nau

Statice
Family: Plumbaginaceae
Genus, species: *Limonium* sp.

The term stance is not recognized botanically as the genus of the large group of everlasting plants that are well known throughout the US and Europe. Taxonomists classify statice under the genus *Limonium*, but floriculturists have commonly called it statice for well over 40 years. The species grown throughout our trade today are used as bedding plants, cut flowers or as perennials. The following types can be used as cuts, either fresh or dried, except for *L. suworowii*, which should be used as a fresh cut flower only. Following is a list of commercially available varieties and their culture.

Limonium sinuatum
Sea lavendar Annual statice

This is probably the most well known member of the statice family. Sinuatum types are best treated as annuals, since they flower the same year they were sown. There's a wide range of colors to pick from: rose, pink, white, purple, blue, lavender and many shades in between. Many growers are also familiar with yellow-flowered sinuatum types such as Gold Coast (actually a sub species of *L. sinuatum*). The height of sinuatum types ranges from 20 to 32 inches, depending on the variety.

Propagation: Sow seed for bedding plants or cut flowers in a peat-lite media; leave seed exposed to light (some growers cover lightly with vermiculite to increase humidity around the seed for more uniform germination). Keep in mind that most types of statice seed is available clean or uncleaned (whole flower heads). Cleaned seed has a higher germination rate and can cost up to 60 percent more. Cleaned seed has about 8,500 to 11,000 seed per ounce, while whole flower heads have 350 to 500 seed per ounce. Seed germinates at 70 F soil temperature from two to 20 days after sowing, depending on age and whether it was cleaned or not. Seedlings can then be transplanted in two to three weeks.

Bedding plants: For growing as bedding plants, transplant seedlings into the final cell pack using one plant per cell. Consider flats that have 32 to 48 cells for green pack sales eight to 10 weeks after sowing. Grow on at 60 to 62 F nights until established (one to two weeks) and then drop the night temperatures gradually to 50 to 55 F nights until sale. It is this cool period for no less than four to five weeks that encourages sinuatum varieties to bloom earlier. Plants will produce a rosette of six to 10 leaves and be about 1 to 2 inches tall at the time of selling—ideal for planting in the garden.

For 4½-inch blooming pots, we have tried one to two plants per pot, but haven't had the uniformity or short crop time that comes with green pack sales. In the home garden, place in a full sun location at 10-inch by 10-inch spacing for flowering plants from July until frost.

Cut flowers: For greenhouse cuts in the Midwest and Northeast, sow seed in January, transplant to a raised greenhouse bed or bench in March or April, for flowering in May. Follow the procedures above for growing starter plants; transplant to the bed or bench six to eight weeks after sowing. Grow on at 60 to 65 F days and 55 F nights. It is our experience that sowings made in the fall in our climate are weak-stemmed and stretch due to lower light conditions. *L. sinuatum* is not photoperiodic, but freely flowers under long days. Sowing earlier than January may not be profitable.

For field growing in the Midwest, follow the procedures above for bedding plants, but transplant to the open field eight to 10 weeks after sowing. Greenhouse sowings in March will produce transplants ready for field planting by late May or June. Plants will flower from July until frost. Spacings should be no less than 10 inches by 10 inches; at a 12-inch by 12-inch spacing, quality increases. Seed may also be sown directly into the field.

For field growing in warm winter areas, successive sowings from July to February every six to eight weeks will provide continuous cropping from December to late June or later depending on the area and climate. Plants are transplanted from plug trays with either one or two plants per plug. Space at 1 foot distance between rows—based on equipment used to plant, cultivate and harvest the crop. In areas where the temperatures seldom get below 50 to 55 F, gibberellic acid is used in the field at 500 ppm when plants are 6 to 8 inches wide to promote earlier flowering.

In general, sinuatum types as cut flowers do not need staking or support unless they're at close spacings.

Statice is considered a medium feeder and will produce inferior cut flowers if given lush conditions. Constant liquid feed at 150 to 200 ppm during the seedling stage and then again in the field as needed. Light sandy soils require more frequent feedings than heavier clay loam soils.

Harvest cuts when the blooms fold outward and are showing color. As a fresh cut, flowers last one to two weeks and will naturally dry in the arrangement. Sinuatum types will last two to three weeks stored fresh at 34 to 36 F. For the best quality stems, use them the first week or so. To dry statice, hang a bunch of seven to 10 stems upside down in a dark, ventilated area for a period of two weeks.

Limonium suworowii
Rat-tail statice, Russian statice

At the height of the cut flower era, long before bedding plants gained popularity, rat-tail statice was popular in northern greenhouses as a fresh cut flower. Unlike its relatives, *L. suworowii* is best used fresh; it doesn't dry well. Rat-tail statice is characterized by light pink blooms on 15- to 20-inch stems. As the name implies, blooms have a tail-like appearance and make excellent cut

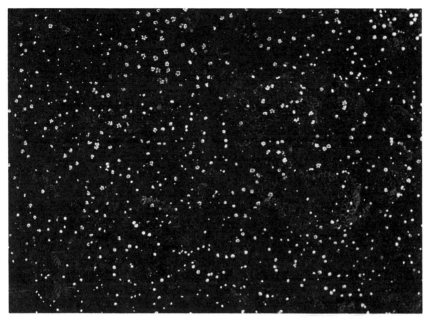
Purple *Limonium sinuatum*

flowers in the winter in a northern greenhouse or in the field in warm-wintered areas. *L. suworowii* has approximately 195,000 seed per ounce (cleaned) and should be sown, germinated and grown under the same procedures outlined above for *L. sinuatum*. The primary difference between culture for the two is timing.

In the Midwest and Northeast, sow seed in September for cut flowers in February. Transplant to 2½-inch pots or directly to the bench. Space in the bench at 8 inches by 8 inches. Grow on at 55 F nights. For growers on the West Coast and in the Deep South, sow at the same time or a little earlier and transplant to the field from pots or as plugs. Growers in warm winter areas will find this crop comes into bloom several weeks earlier due to higher light as well.

In dark winter areas, try growing this crop in 5- or 6-inch pots to flower. As plants become root-bound, they tend to flower easier. Incandescent lights help flowering too.

Limonium perezii
Seafoam statice

Often sold as a perennial in the Midwest, *L. perezii* will be short-lived at best. In our Chicago location, we have had excellent results treating this as a bedding plant and making no promises it will live from year to year. *L. perezii* is characterized by large blue flower heads on upright stems. Follow the same procedures for sowing *L. sinuatum*. Seafoam statice has 24,000 seed per ounce (cleaned).

Sowings in mid-March produces green packs for sale by June 1. Sow in early to mid-February. Transplant one or two plants per 4-inch pot and sell green in May. Sold in 4- or 5-inch pots, *L. perezii* will bloom in the garden in late June or July. In areas with severe winters, the variety will grow to 2 feet at best. In warm winter areas the variety is a perennial and will reach a height of 3 feet.

Limonium

Limonium latifolia

L. latifolia is a true perennial statice that will flower the second season from seed. Prominent, strap-shaped leaves form a large rosette (15 to 18 inches across) once established in the field. Flowers are lavender-blue in large panicles that are airy and open. Flowers are smaller than the other statices mentioned.

Germinate and grow on under the same conditions as above, but remember that *L. latifolia* will not bloom the same season from seed. Also, it will require approximately 10 to 12 weeks to be salable as a 2¼-inch plant from a spring sowing.

Once they have begun to flower, cuts can be harvested for several years.

Goniolimon tatarica
German statice (*Limonium tataricum*)

This plant is a horror story when it comes to taxonomy. Every source taxonomically gives it a different edge. Though related to limonium, German statice is slightly different morphologically. There is one major difference to be aware of if you are going to grow this variety: Crop time. German statice is a slow grower.

A striking plant, German statice has silvery to dusty grey bract color with light lavendar blue flowers. As cuts dry, flowers drop and the bracts become the ornament. Growing to about 20 inches, German statice requires two years to bloom from seed. It is an excellent perennial for the North and we have had plants in our gardens here for many years.

Germination and growing on is the same for German statice as for the other statices mentioned. There are approximately 22,000 seed per ounce (cleaned). The first season after sowing, plants will stay small and will remain in a rosette. During the second season, plants send up a flower stalk in late April or May, but will not bloom until mid-June. Plants are deceiving—they bud up and then wait for up to three weeks before they start to open. Once open, flowers are in color for several weeks. Cut flowers when the plants show the amount of color you want; once cut, flowers do not open any further.

To grow German statice in packs or pots, allow at least 15 to 18 weeks to produce a 4- to 5-inch rosette. If plants are to be sold the same season as they have been sown, then sell in packs only. Plants are not aggressive enough to fill out a pot unless sown early; this may not be practical when German statice is grown as a perennial. Sowings made in June or July are potted into 4-inch pots or quart containers by September and overwintered in a coldframe. These plants will bloom the following year, and will probably be budded at the time they're sold (late April or May).

As a cut flower, consider intercropping German statice in a block with other crops. Since plants flower for only a very short period, the field would remain idle for almost 10 to 11 months out of the year in the Midwest.

"

Jim Nau is trials and product development manager for seed at Ball Seed Co., West Chicago, Illinois.

Mums, see Chrysanthemums

Narcissus, Daffodils

Culture Notes

September 1986

by Teresa Aimone

Daffodil
Family: Amaryllidaceae
Genus, species: *Narcissus* sp.

Daffodils are true bulbs. The bulbs consist of a slender piece of stem tissue surrounded by numerous fleshy leaves. The leaves contain the stored food material; the stem contains the flower infloresence. The flower buds of this genus form during the spring and summer of the previous year—usually when flowering in the field is nearly over, but just before bulbs are harvested for next year's sales. In contrast, hyacinths and tulips form their flowers at the end of the growing period or after they have been placed in storage.

Growing techniques

Bulbs are cooled to promote stem elongation—not to develop flowers. Cooling temperatures between 35 to 50 F cause chemical changes within the bulbs; when temperatures rise following cooling, bulbs begin growing rapidly (forcing stage). During cooling, starch contained in the bulbs changes to sugar; this sugar is used in the formation of proteins and other cell-building products needed for growth activity during the warmer forcing temperatures in the greenhouse. If bulbs are overcooled, they will use all their available stored food and produce unsalable plants.

Cool temperatures are not needed for root development. Temperatures below 35 F will impair root growth. Root development must be complete prior to the onset of freezing weather if bulbs are being cooled in outdoor frames. Root development should be complete in four to five weeks. Medium should be kept moist at all times during cooling. If rainfall is not adequate for outdoor-cooled bulbs, provide periodic watering.

Planting: When bulbs arrive, open the boxes and ventilate the bulbs. Inspect the bulbs for disease. For best results, plant in bulb pans or azalea pots. Pots should be sterilized prior to planting. Either clay or plastic pots can be used. If you're growing the bulbs for cut flowers, plant them in wooden flats—14 inches by 16 inches by 4 inches. The wood should be treated with copper naphthenate prior to planting. For good drainage, the bottom boards of the flats should be spaced ¼ inch apart.

Medium: Soil medium should be well-aerated, yet have the capacity to retain sufficient moisture for good growth. Sterilize the medium prior to planting. A medium which drains too readily (such as pure sand), or a heavy medium (such as one with a clay base) should be avoided. Loam soils with perlite, vermiculite, or calcinated clay in a 1:1:1 ratio work well. Maintain a pH of 6.0 to 7.0. The soil temperature, at the time of planting, should be approximately 50 to 65 F.

Fertilization: Daffodils don't necessarily need to be fertilized during their production cycle, since the bulbs themselves contain enough stored food to provide proper nutritional requirements.

Production temperatures: No matter when the bulbs are planted, they should be given temperatures of 48 F until December 1 to 5; 41 F until January 1 to 5, then finished at 32 to 35 F. The change to different temperatures depends on plant development. A temperature of 48 F is for root development—the temperature should be maintained until roots emerge from the bottom of the container. A temperature of 41 F is necessary for shoot growth. Maintain temperatures of 41 F until shoots are approximately 2 inches long. Temperatures of 32 to 35 F hold the shoot growth until plants are brought into the greenhouse for forcing.

Greenhouse forcing temperatures for pot daffodils are 60 to 62 F. Optimum stem length for pot daffodils is 10 to 14 inches. For cut daffodils, greenhouse forcing temperatures should be maintained at 55 F. This lower temperature will induce the longer stems desirable in cut flowers. Avoid temperatures above 60 F, since weak stems will result. Store both cuts and pots at 32 to 35 F no longer than one week. Harvest both at the "gooseneck" stage of development. A desirable stem length for cuts is 14 inches.

Timing: Potting can be done anytime from October 1 to December 1, depending on flowering date. For Christmas flowering, plant October 1; for Valentine's Day, plant in mid-October, and for a March and April flowering, plant in mid-November. Bulbs forced for earlier flowering need to undergo a precooling treatment. Precooled bulbs should arrive at the end of August and be placed in storage at 48 F no later than August 31. Do not plant the bulbs until October 1 for December flowering.

Height control: Daffodils respond to Florel (ethephon). Apply 1,000 ppm (3.2 ounces per gallon) to 2,000 ppm (6.4 ounces per gallon) when the leaves and/or floral stalk are 4 to 5 inches tall. Apply to dry foliage, and don't get the foliage wet for 12 hours after treatment. Don't apply when flower buds are visible, and be sure to trial a small sampling of plants before using Florel on the entire crop. If a second application is required, apply two to three days after the first treatment. Apply Florel in a well-ventilated, 60 to 63 F greenhouse.

Diseases: Most disease disorders occurring on bulb crops are due to diseases present on the bulbs when they arrive. Check the bulbs carefully after you open the box. A common disorder of daffodils is Fusarium basal rot. This originates in the production fields. An identifying characteristic of this disease is soft tissues, particularly on the bottom of the bulb. If more than 10 percent of the shipment has this disorder, return the shipment.

"

Teresa Aimone, former editor of GrowerTalks *magazine, is with Sluis and Groot, Fort Wayne, Indiana.*

Nephrolepis, Boston fern

Culture notes

February 1990

by John Erwin

Boston Fern
Family: Polypodiaceae
Genus and species: *Nephrolepis exaltata* 'Bostoniensis'

Boston fern is a popular pot plant or hanging basket. Often it's grown as a secondary crop in a greenhouse where environmental conditions are optimum for development of the primary crop, which is growing below fern baskets. As a result, Boston ferns often don't grow as rapidly as possible. This information may help when deciding which crops to grow with ferns to optimize growth of both crops.
• **Background.** Boston fern is a cultivar of the sword fern, *Nephrolepis exaltata*, a terrestrial or epiphytic fern native to the American tropics and subtropics. In 1894 a Philadelphia grower shipped 50,000 sword ferns to a Boston distributor. The plants were different from the traditional *Nephrolepis exaltata* and apparently arose from a sport. The new cultivar was named *Nephrolepis exaltata* cv. Bostoniensis, or the Boston fern.

There are a tremendous number of Boston fern cultivars grown commercially today. Most cultivars haven't arisen from hybridization, but from mutation of runners.

Propagation
Most growers receive fern plugs propagated using tissue culture. Typically, plants propagated through tissue culture are sold in a 72 plug tray. Often growers in the North receive plug trays in late spring for fall and winter sales.

Plants are grown in 4-inch standard, 6-inch azalea, 6-inch hanging baskets or 8-inch hanging baskets. Pot one plug per 4-inch pot, two plugs per 6-inch pot or basket and four plugs per 8-inch basket. The number of plugs per pot can vary based on desired production time. Typical production time for the plant numbers and pots described above is 18 to 20 weeks.

Growing on
• **Temperature.** Early research showed that frond length and overall plant quality decrease as night temperature decreases from 70 to 60 F. Day temperatures above 100 F also decrease plant quality at market.

Recent research shows that Boston fern growth is primarily influenced by average daily temperatures when plants are grown between 50 and 86 F. Frond length, frond number and leaflet number all increase as the average daily temperature increases from 50 to 76 F. Interestingly, as average daily temperature increases above 76 F, frond number decreases, although frond length and leaflet number still increase.

Average daily temperature also affects total plant dry weight. High plant dry weights usually mean higher quality. Specifically, fern plant weight is greatest when average daily temperatures are approximately 72 F.

Grow ferns at 72 F, keeping day and night temperatures both as close to 72 F as possible. Allowing either day temperature or night temperature to rise above 80 F reduces the number of fronds per plant.

- **Photoperiod.** Boston ferns grown with a nine-hour photoperiod have more but shorter fronds than ferns grown with a nine-hour photoperiod plus a four-hour night interruption from 10 p.m. to 2 a.m. Therefore, to get longer fronds on plants grown during winter, consider using "mum lighting" with incandescent lighting at an intensity of 10 footcandles to increase frond length and leaflet number.
- **Nutrition.** The nutritional requirements for Boston fern are rather modest. Continuous liquid feeding with a 200-0-200 parts per million mix should be more than adequate for growth.

Media pH has a dramatic effect on fern growth. The optimum pH for Roosevelt fern is between four and five. Growing ferns with a pH from 6.2 to 7.0 reduces growth by 25 to 50 percent! Pay close attention to media pH; keep it below or around 6.0, if possible.

- **Watering and media.** Many difficulties in growing Boston fern arise from too little or too much water. Allow Boston ferns to dry slightly between waterings. If plants are grown wet, root rot problems may develop.

Because of difficulties with root rot, be certain to use a well aerated soil mix. Good drainage is essential. Do not use a mix containing any soil under any circumstances. If a root rot problem should develop, apply what is registered for Pythium and Rhizoctonia control on ferns in your state.

- **Pests.** Insect pests are often not a problem with ferns. Mealybugs can become a problem in the crown. The easiest way to eliminate this pest is to apply a systemic pesticide.
- **Acclimatization.** Ideally, all fern plants should be acclimated prior to sale. Acclimatization ensures better growth under low light conditions in a home. Boston fern benefits greatly from a single week of acclimation under low light conditions (less than 750 footcandles). Acclimation of Boston fern for longer than one week doesn't significantly prolong postharvest life.

Cultivars

Many cultivars are available. More compact cultivars are grown in pots, whereas larger fronded cultivars are grown in hanging baskets. The standard Boston fern and the cultivar Napa Valley sell well in hanging baskets. Dallas Jewel and Compacta sell well as 4- and 6-inch pots.

For best growing success

Boston fern quality is best and frond production rate most rapid when plants are grown with day and night temperatures between 72 F and 74 F. A crop needing high temperatures that would grow well with Boston fern baskets would be New Guinea impatiens.

Use photoperiod control to increase or decrease frond length; long days lengthen fronds, short days shorten fronds. Be careful when using incandescent lighting to give long days if ferns are growing above another crop, as this may stimulate elongation in the lower crop.

99

John Erwin is assistant professor, floriculture physiologist and extension specialist, University of Minnesota, St. Paul.

Nerine

Culture notes

October 1985

by Teresa Aimone

Nerine
Family: Amaryllidaceae
Genus, species: *Nerine bowdenii*

With the increasing consumer awareness of new and exciting cut flowers, it's time that growers here take advantage of the situation and grow these items for this eager U.S. market. The nerine is one cut flower that fits this description. These exotic flowers come in white, pink, yellow, orange, red and bicolors. And, with both proper cultural techniques and post-harvest care, nerines can provide a long vase-life for consumers.

Consumers love the exotic look of this uncommonly seen cut flower, and cultural requirements are such that nerines can be grown in many parts of the U.S. They are best suited for production in the more temperate regions.

Planting

Nerines, as members of the Amaryllidaceae family, grow from bulbs. Bulb sizes start at 12 cemtimeters; depending on the size of the bulb, they should be planted 12 to 15 per square foot.

Nerine

Recommendations for medium suggest that all soil type mixes are suitable as long as good aeration and proper drainage are provided. Before planting the bulbs, dip them in a fungicide solution. Suitable chemicals are Benlate, Topsin M, Mertect and Tursan 1991. Check labels for recommended rates. Plant the bulbs so just the nose of the bulb shows above the soil line.

Ferilization

Information on this cultural point suggests incorporating 1$\frac{1}{3}$ square yards of composted manure per 100 square feet prior to planting; approximately two weeks prior to flowering, fertilize with 6$\frac{1}{2}$ pounds 12-10-18 per 100 square feet.

Greenhouse cultivation

Nerines are a long-term crop; flowering occurs approximately 189 to 198 days after planting. Research has been done to try and cut the crop time down; and though one experiment cut three weeks off the growing schedule, this "quick-cropping" reduced flowering by two-thirds. All this work does indicate, however, that the length of time from planting to flowering is governed by temperatures; adjustments in these temperatures, in turn, are determined by the time of planting. Table 1 lists planting and respective flowering times.

If bulbs are planted in January and February, maintain a 56 F night temperature until April. Raise temperatures to 61 to 65 F night after the beginning of April. Higher night temperatures should be avoided as they can cause bud blast and encourage basal rot brought on by Fusarium. Higher temperatures will also speed rooting time, but won't speed flowering. Lower temperatures will only increase crop time.

Like all bulb-grown crops, keep the medium uniformly moist; don't let plants dry excessively or flowering will be delayed. Soggy soil can result in both root and bulb rot.

In the greenhouse, final stem length for nerines will probably be around 24 inches.

Planting time	Flowering time
Mid-January	End of August
February	September
May	November
June	December
July	January

Table 1. Planting and flowering times for indoor production of *Nerine bowdenii*.

Outdoor cultivation

If nerines are to be produced outdoors, plant them from the end of March through April for flowering in September and October. Final height at flowering for outdoor-grown nerines will probably be about 12 inches. Follow all other cultural requirements as closely as you can under outdoor conditions.

Harvesting and storing

Stems should be cut when the first floret has colored up and is ready to open. Flowers may be stored at 36 to 41 F for no longer than seven days. Temperatures lower than 36 F can cause the flowers to turn blue.

Re-using bulbs

Nerine bulbs can be used for more than one flowering season. About two weeks after flowering, bulbs can be removed from the medium. After the bulbs have air-dried, they can be stored at 36 F for up to six months. Do not replant the bulbs unless they have been stored at the above-mentioned temperatures for at least three months. (Nerine bulbs can be stored at temperatures up to 50 F.) Do not let the bulbs dry out completely or they can't be reused. If bulbs are grown outdoors they must be brought in during the winter.

99

Teresa Aimone, former editor of GrowerTalks *magazine, is with Sluis and Groot, Fort Wayne, Indiana.*

Information for this article was supplied by both the Netherlands Flower Bulb Information Center and Dr. August De Hertogh, North Carolina State University, Raleigh.

Ocimum, Basil

Culture notes

March 1988

by Thomas De Baggio

Basil
Family: Labiataceae
Genus, species: *Ocimum basilicum*

Basil is the king of the summer garden and the most popular herb plant in America.

A member of the mint family, a group of over 3,500 species which form the backbone of culinary herbs, basil is native to Europe and the tropics.

What's responsible for its popularity in the U.S.? A dish called pesto, a concoction of olive oil, pine nuts and hard-grating cheeses that was created in Genoa, the Italian port city. Traditionally paired with pasta, it has achieved notoriety on everything from potatoes to soup, from chicken to fish. It is a keystone for the food-as-entertainment generation of gardeners who have made herbs a popular commodity at the garden center.

In times past, all sorts of magical properties were conferred on basil. But today most of the magic is finding its way into growers, cash registers. Grown properly this small annual can be one of the most profitable plants any grower could dream of—33 days from sow to sell for a premium, potted plant that retails from about 75¢ to well over $1 each. That means a yield of more than $20 per square foot every 33 days for the mid-priced retail-grower and about half that for the wholesaler—not an atypical revenue picture for the grower of potted herb plants.

Propagation: Transplant to sale in 11 days

Typically basil is grown from seed. There are about 18,000 seed per ounce of basil with a viability of approximately 60 percent.

I start my seedlings the old-fashioned way in a 1020 plastic tray. (The seed is large enough and smooth enough to use in an automatic seeder. I have nothing against plugs, my operation just isn't large enough to warrant such an investment.)

Sow basil seed around the first week in March. This will produce salable plants by mid-April, too early to transplant into gardens in zone 7, (basil needs warm ground and should be planted at the same time eggplant goes in). This is about as long as I can hold my customers at bay. Their craving for basil often defies gardening logic.

Ocimum

Sow seed thickly with seed touching and on top of each other. The rows may be covered or left uncovered. Use a coarse mix for herb culture—I use Pro-Mix BX as both the germination and growing medium.

Flats are covered with plastic or a humidity dome and placed in a warm room where temperatures are maintained at 70 F. In four days the seed is germinating and the flats are moved under fluorescent lights. Keep lights about 5 inches from the top of the flats; 16-hour days are provided with temperatures varying from 65 to 75 F. Begin weekly fertilization with 20-19-18 Peat-Lite Special seven days after germination. This continues until sale.

In 18 days seedlings are ready for transplanting into 2½-inch pots in the greenhouse. I prefer the deep Nu-Pot since it provides a longer shelf-life for these rapid growers.

The key to my fast-cropping method is in the transplanting. I do not transplant single seedlings. I use clumps of four or five seedlings. This lessens transplant shock and produces the effect of a branched plant from the start. The seedlings are placed in a dibbled hole so that only the true leaves are above the growing medium. This is very deep transplanting and is necessary to keep the plants stocky.

Plants are not meant to be separated before transplanting to the garden. The end result: 40 quarts or more of fresh leaves may be harvested from the plants by the gardener during the summer growing season.

Fungus disease is the chief problem after transplanting. Since no chemicals are licensed for basil (and most herb plant purchasers abhor chemical sprays anyway), prevention is the best choice. Keep temperatures in the upper 60s to mid-70s nights and keep the fan-jet working all the time. Cooler temperatures during the day are important to produce stocky plants and prevent disease, so ventilate during the days.

Whitefly is the chief insect problem, although spider mites can also damage plants once temperatures begin to warm in early summer. They are easily controlled with weekly applications of Safer Insecticidal Soap, a contact spray that works wonders when properly sprayed up through the plant so the undersides of leaves are coated where the whiteflies and spider mites live.

Eleven days after transplanting, basil is ready to sell.

Cultivars

Basil is in a realm of its own, full of subtle flavor, height, leaf-size and color differences. In the '80s alone, at least six new basil varieties have come on the market in the U.S. and more are in the wings. And that doesn't include new arrivals from Europe, Asia and Africa.

All the variety tends to obscure the fact that most buyers are looking for "the pesto basil, please." Although most U.S. seed catalogues offer something called "sweet basil" (*Ocimum basilicum*), the true Italian basil used for pesto is a variety called Genoa Green. It resembles sweet basil in appearance, but not flavor, being less cloyingly sweet. It is sufficiently better that it has been worthwhile to import my own seed for the last decade. I've known customers to drive hundreds of miles for a dozen plants of Genoa Green basil. (It is now available in the U.S. from Johnny's Selected Seeds, Foss-Hill Road, Albion, Maine 04910.)

The point is that with so much generic basil floating around garden centers, a named variety gives customers something familiar and predictable to hold onto year after year. Spicy Globe basil has become a favorite for dwarf basil hedges and a delightful potted plant for windowsills, patios, porches or windowboxes. Green Bouquet is a selection from what we have called for the last 300 or more years *Ocimum basilicum Minimum* or bush basil.

While a good cook would never think of making pesto with one of the colorful purple-leafed varieties, they are delicious in salads and a wonderful, spicy, basil vinegar can be made from them. It is also, along with the dwarf bush types, becoming an essential part of the bedding garden where its color and form provide texture and interest.

There are also varieties with huge, puckered leaves (Purple Ruffles, a 1987 All-America Selections winner) as well as basils with lemon-scented leaves licorice-flavored leaves and leaves with a cinnamon taste. There is a basil with a camphor scent and some sweet basils from Asia with flavors that withstand the heat of frying.

Marketing tips

Pull out the stops for marketing basil and other herbs. The small plants are fine for early in the season, but 6-inch and 1-gallon pots sell well later when the basil lover just can't stand it anymore and wants a plant to eat tonight with some leftover for the garden tomorrow.

While basil is probably too large a plant for a small windowbox collection of herbs (in fact, windowboxes usually don't last longer than a month for the customer because the plants grow so fast they become pot-bound), it can be grown in grow-bags for the customer with a craving but without any space or in tubs for the patio or balcony.

It's unlikely that you'll have to try hard to sell basil. The problem is having enough of it to satisfy demand.

"

Thomas De Baggio, a former director of the International Herb Growers and Marketers Association, is owner of Earthworks Herb Garden Nursery in Arlington, Virginia.

Otacanthus

Culture Notes

April 1986

by Teresa Aimone

Otacanthus
Family: Scrophulariaceae

The arrival of the new flowering pot plant and cut flower, otacanthus, is right in step with the times. Both growers and consumers, alike, are anxious to try new items, and here's something that's ready to fill that bill. In addition to their newness, otacanthus also has the added bonus of blue flowers—and there aren't too many of those around.

The full sun requirement of otacanthus also makes it especially attractive for 4-inch bedding plant sales. Otacanthus can be used in outdoor plantings of annual flower beds and borders, patio containers, planters, and urns.

Otacanthus is relatively new to this country. They are being researched in Denmark for use as a viable production plant there. Since otacanthus is so new to this country, they are currently available in rooted cutting stage.

Medium

Best results for otacanthus are received when plants are grown in a well-drained, porous medium. Recommendations suggest that plants be grown in a mix containing 50 percent peat moss and 50 percent aggregate (perlite, polystyrene beads, etc.). You may wish to try growing otacanthus in a commercial peat-lite mix if that is what you're currently using.

pH

For best results, maintain a pH in the synthetic medium of 6.0 to 6.5. If pH is too alkaline and rises to 7.0 or above, the foliage may yellow.

Temperatures

Maintain a night temperature of 62 to 65 F. Hold daytime temperatures between 70 and 75 F; higher temperatures will cause plants to stretch.

If the crop of otacanthus is ahead of schedule, the grower can delay both the growth and flowering if the night temperatures are reduced to 50 F. This procedure will work only if the flower buds are visible. When grown outdoors, high temperatures improve flowering.

Otacanthus

CO_2
Enriching the greenhouse environment with CO_2 injection will enhance the growth of otacanthus under low light conditions. To obtain the best results, maintain CO_2 levels of 1,000 to 1,500 ppm in the greenhouse.

Light
Otacanthus are high-light plants. They should receive high light in the greenhouse; this will promote flowering and reduce plant stretch. Otacanthus like the same conditions when planted outdoors—plant them in full sun. Low light levels will reduce flowering and cause plants to stretch.

Irrigation
As mentioned earlier, otacanthus like a well-drained, porous growing medium. Keep the medium moist, but not soggy. If otacanthus is allowed to wilt, it can tolerate this without dropping flowers. However, if allowed to wilt under bright light, the plants may burn.

Nutrition
Otacanthus is not a heavy feeder. Maintain a constant liquid feed level of 150 ppm N, 100 ppm P_2O_5 and 150 ppm K_2O. Lower levels will cause the foliage to become chlorotic (yellow).

Spacing
Otacanthus likes to grow tall, so give them adequate spacing on the bench or in outdoor production. If the finished plants are to be grown in 4-inch pots, provide a 6-inch by 6-inch or a 7-inch by 7-inch spacing. If 5½-inch pots are used for the finishing containers, use a 9-inch by 9-inch spacing.

Scheduling

Timing, of course, depends on season and locale. For that reason, scheduling of otacanthus can vary between eight and 12 weeks to produce a 4- or 5½-inch pot from a 72-cell pack liner. When plants are received, immediately unpack them, pot them up, and pinch the plants. There should be two sets of leaves left after the pinch.

Growth regulators

Since otacanthus is so new, the chemicals that will produce the desired results aren't labeled for otacanthus. Therefore, these are chemicals that have been used in testing situations.

A 0.5 percent concentration of B-Nine should be applied to plants one week after pinch; a second application, at the same concentration, should be applied 10 days later, and a third application (at 0.5 percent) applied 10 days after the second to promote compact plants. As we said at the beginning of the article, otacanthus can also be used as a cut flower; they do have a tendency to elongate. That's why they need so much growth retardant. Plants that receive the B-Nine treatment will flower approximately one week earlier than plants that do not receive the treatment

Literature states that Cycocel can be effective in height control, but it will cause yellowing of the foliage. The above treatment is the same for 5½-inch pots. Two plants per pot are recommended for this size.

Diseases

Botrytis can be a problem if plants are grown close together during periods of cloudy weather. There are chemicals that have been said to control Botrytis on otacanthus, but there is no label registration for them. The chemicals are Chipco, Termil, and Benlate.

Insects

Otacanthus generally do not have a problem with insects. Growers may occasionally have a little problem with aphids or whiteflies, though. Again, there are chemicals that are said to control these insects on otacanthus, but there are no label registrations at present. Those chemicals are Orthene for aphids and Vydate for whiteflies.

"

Teresa Aimone, former editor of GrowerTalks *magazine, is with Sluis and Groot, Fort Wayne, Indiana.*

Culture notes

December 1989

by Lin Saussy Wiles

X Pardancanda (Candy Lily)
Family: Iridaceae
Genus, species: X *Pardancanda Norrisii*

Pardancanda, or candy lily, is among the few multi-purpose plants that can be grown as perennials, in beds, in wild gardens, in borders or in pots with equal success. The dwarf forms (14 to 19 inches tall) are especially adapted to use as pot plants. The cut flower industry will be pleased to find a plant with characteristics similar to alstroemeria, but much easier to grow. Individual flowers last one day, but flower buds continue to open for two weeks or more per stalk.

Propagation
- **Clones.** Divide plants in the fall to reproduce choice individuals. Divide after the old fans turn brown and new young fans appear at the base of the plant. Each division should have a minimum of two to three fans present. Some plants produce enough young fans the first fall to yield five or more divisions. Use a sterile, sharp knife to separate the plant and a fungicide on the cut surface prior to repotting.
- **Seed.** Pardancanda seed are quite large (1,100 seeds per ounce) and can be sown in an average sowing medium. Seed sown in early February will frequently flower in August of the same year. Get best results by sowing in November or December for flowers the following year. Cover seed to the depth of the seed. Bottom heat is not necessary. Seeds germinate in three to seven weeks. Sowing date has little effect on time of flowering within the November to February time frame.

Growing on
- **Transplanting.** Transplant plants about 10 weeks from sowing when plants are 2 to 4 inches tall. Dwarf plants will be smaller than semi-dwarf or standard plants even as seedlings. Transplant into 2½-inch pots using a well-drained peat-lite medium. Seedlings are fan-shaped and grow rapidly.

Transplant to 4-inch pots in four weeks. They can be transplanted again to 6-inch pots in four weeks. Plants are adaptable to being root-bound and can be held in 2½-inch pots through flowering, although severely root-bound plants will not flower as well as those potted up to larger containers.

Pardancanda

- **Lighting.** Plants have a high light requirement. Full sunlight produces best flowering and plant quality. Lower light levels cause stem length to increase somewhat within each height class—dwarf (14 to 19 inches), semi-dwarf (20 to 25 inches) and standard (26 to 45 inches).
- **Temperature.** These plants thrive in high temperatures. Minimum temperature tolerance during the seedling stage is around 50 degrees F, but mature plants are winter hardy perennials.
- **Fertilization.** After transplanting into 2½-inch pots, begin constant feed using 20-20-20 at 75 to 100 parts per million nitrogen at every watering. Continue constant feed after transplanting into 4-inch pots. When plants are ready for 6-inch pots, continue constant feed after transplanting or add 1 teaspoon Osmacote per pot.
- **Watering.** Keep plants evenly moist until they are large enough for 6-inch pots, then occasional drying down will not be greatly detrimental. Mature plants are exceptionally drought-tolerant, but best flowering occurs with good irrigation.
- **Insects.** Insect problems seldom affect flowering or general appearance to any extent. Treat spider mites using Pentac. Thrips and mealybugs sometimes occur on mature plants. Aphids occasionally occur on seedlings.

Pardancanda

- **Disease.** When Botrytis occurs on seedlings or around flowers on mature plants, Benlate is an effective treatment.
- **No growth regulators,** pinching or other special requirements are needed.
- **Cultivars.** Candy Lily is a mixture of a number of dwarf, semi-dwarf and standard lines bred by Park Seed. Plants are highly drought tolerant and produce flowers daily over several weeks the first year and over several months in subsequent years.

"

Lin Saussy Wiles was a plant breeder for Park Seed, Greenwood, South Carolina when this article appeared.

Pelargonium, Regal geraniums

Culture Notes

October 1987

by Ron Adams

Regal geraniums
(Martha Washington geraniums)
Genus, species: *Pelargonium* x *domesticum*

Regal geraniums have been under cultivation and development for over 200 years. They are popular flowering plants with colorful marked petals that resemble azaleas. The popularity of Regal geraniums has been increasing and would increase more if their outdoor garden performance was not regionally limited because of cool growing requirements. Historically they have been considered a florist crop. Researchers hope to develop strains suitable for bedding plants, with continuous summer flowering. Kenneth Post described the following cultural practices for Regal geraniums in 1949:

1. Cuttings were taken in August and September and rooted.
2. Plants were grown at low temperature, approximately 50 F through fall and winter. This led to flowering in the spring.

Since Kenneth published these cultural practices a lot of research has gone into determining precise flowering mechanisms.

Today, you can buy initiated established rooted cuttings or liners for finishing in 4- or 6-inch pots that will flower in only eight to 10 weeks. Night temperatures and light are the primary factors for controlling flowering. Because cool night temperatures are required, outdoor performance varies regionally. Regals are best grown outdoors in the Northeast and Northwest where night temperatures are lower. Warm night temperatures in other parts

Pelargonium

of the country cause the plant to become vegetative. Even if you are able to bring them into flower, they will not continue to flower all summer long.

Propagation

Let's assume that we are going to produce a Mother's Day crop for 1988 from our own stock plants.
- Pot well-rooted cuttings or liners in 6-inch pots on October 12, 1987.
- Take cuttings on January 4, 1988.
- Begin cold treatment on February 1.
- Pot initiated cuttings in 6-inch pots on February 29.
- Flowering finished pots will be ready May 2.

Crop time is approximately 29 weeks from established stock plant rooted cuttings.

Stock plants: It takes about three months to produce a well-branched stock plant. Start from a rooted cutting and grow on at 60 to 65 F night temperature. Fertilize with a 200 ppm to 250 ppm N constant liquid feed (15-16-17 or 20-10-20 peat-lite), switch ppm every fourth watering to calcium nitrate at a 200 ppm N. Grow in full sun. Properly pinching stock plants is required to have sufficient breaks to create cuttings required (one or possibly two pinches). Maintain CO_2 levels at 1,000 ppm to 1,500 ppm.

Cuttings: Stock plants should be watered the day before taking cuttings. Take cuttings that are approximately 2½ inches long and have two mature leaves. Cuttings can be snapped off or taken with a knife an inch below the leaves. Make sure that the knife and the propagator's hands are sterile. Wash hands often with soap and water while taking cuttings to reduce the chance of infections.

Pelargonium

Unrooted cuttings should be stuck in a sterile, well-drained media. Container size varies from 2¼ to 3 inches depending on how long the cutting will be grown before potting. Soil temperature for rooting should be 70 to 75 F. Place cuttings under mist for seven to 10 days. Mist timing should be seven to 10 seconds every 10 minutes when cuttings are first stuck; reduce frequency as the cuttings begin to root. Cuttings will be rooted in 14 days, but should be given an additional two to three weeks at 70 to 75 F to establish. If you are finishing a pinched plant, pinch one week after rooting, while plants are still in the propagation area. This will give the cutting time to develop new growth before floral initiation.

Growing on

Regal geraniums require four to five weeks of temperatures at 48 to 50 F to initiate flower buds. Initiation is easily accomplished in the greenhouse in full sun. Temperatures can go as low as 35 F, but not higher than 50 F during the four-week induction period.

Finishing: After cuttings are initiated, or if you have bought precooled cuttings, plants are ready to be potted and forced. Plant cuttings slightly deeper than the original soil line. Most Regals are grown one pinched cutting per 6-inch pot.

Photoperiod control is critical for floral development. Grow with long days (16 hours at 10 footcandles). Many owners use a night break of 10 p.m. to 2 a.m., beginning when cuttings are potted until the flower buds show color. Calculate 5 watts per square foot of incandescent light for night break or extended daylength.

Grow in a well-drained media but do not allow plants to dry out during production. Soil pH should be 5.5 to 6.0. Apply constant feed at 200 ppm N (20-10-20 or 15-16-17) using a peat-lite formula if growing in a soilless mix. Rotate every fourth watering with calcium nitrate as with the stock plants.

Do not apply supplemental CO_2 during finishing since it will cause rapid vegetative growth.

Final spacing should be 12 inches by 12 inches. Temperatures should be 58 F nights, 68 F days.

Insects and diseases: Whitefly control is critical in Regal production. Regals do not like systemic insecticides such as Temik. Consult your county extension advisor for recommendations on whitefly control.

Regals are susceptible to Botrytis. A regular fungicide program should be maintained during finishing and stock plant production using Chipco 26019, Benlate or Ornalin, and Exotherm Termil. Exotherm Termil should not be used after flowers are open.

Height control: With Regal geraniums, the higher the light intensity the shorter the plants. Some growers report height control using Cycocel at a 3,000 ppm foliar spray rate. Split applications to reduce foliar burn, making the first application 17 days after potting and the second application seven to 10 days later. Apply Cycocel at a rate of 1 gallon of spray for 200 square feet. Additional Cycocel applications will depend on growing conditions and market demands.

Postharvest: Some growers have had excellent results reducing bud shatter caused by a buildup of ethylene in the shipping container with silver thiosulphate applied at the visible bud stage.

"

Ron Adams is technical services manager with Ball Seed Co., West Chicago, Illinois.

Pelargonium, Zonal geraniums

Culture notes

March 1989

by Joe O'Donovan

"Fast cropping geraniums is state-of-the-art production that provides customer satisfaction."

Fast cropping provides the key to profitability and success with geraniums. Fast cropping has become synonymous with the production of 4- or 4½-inch pots, but has application in all phases of geranium production, including stock, propagation and finishing.

Just what is fast cropping? It's state-of-the-art production. As originally defined, the technology revolves around four points: quality stock, media, fertilizers and temperature control.

- **Quality stock** is truly the cornerstone, focusing on cleanliness, as well as the basic genetic trait. Cleanliness is the first order of discussion, for **without clean stock the concept of fast cropping falls apart**! The heavies of yesteryear are still the heavies of today—sort of a Geranium Age Xanthomonas, Pseudomonas and Verticillium. Stock that is infested with any of these pathogens just will **not** work in a fast crop system. Stock that is problem-free makes the system perform like clockwork.

Virus problems also have an impact on quality. Most of the virus problems prevalent in modern day geranium crops result in the production of "scratch and dent specials"—product that has fewer flowers, smaller flowers, marked foliage and possibly less vigor.

- **Variety selection** is also a cornerstone in the fast cropping system. The task has become more complex in recent years, and consumer demand has changed the market mix from red and others to one driven by colors.

New varieties have also changed the nature of the art of production, adding unique shades and tones, earlier flowering, more uniform flower response and more compact habit.

- **A peat based medium** helps you get the most out of fast cropping. Adjust the pH to 6.0 to 6.5 for zonals and slightly lower for ivy leaf varieties. The mix should be pasteurized, well-drained, low in soluble salts and able to exchange nutrients (high CEC).

- **Balanced nutrition** is the key to successful production. Geraniums enjoy relatively high levels of fertility, especially phosphorus and magnesium. Use a constant fertility program of 250 to 350 parts per million (ppm) N once the root system is established. Be careful to contain electrical conductivity in the range of 1.2 to 1.7. This program can be supplemented on a biweekly basis with 1 to 2 pounds per 100 gallons of magnesium sulfate (Epsom salts) to maintain adequate magnesium levels.

- **Accurate temperature control** is a must. Characteristically, geraniums are the crop that really respond to the old adage, "warm feet, cool tops."

Pelargonium

Especially in the initial stages, warm feet help establish an active root system that sets the pace for a successful crop.

During this initial stage, root zone temperatures of 70 to 75 F are imperative. Once the root zone is established with active, ample root growth at the extremities of the soil ball, reduce the medium temperature to 65 to 70 F. A day/night temperature differential of 5 F is necessary.

Split night temperatures can be used without extending the crop cycle. In doing so, make the initial night temperature reduction at sundown, lowering the air temperature by 5 F below daytime "heat-to" temperatures.

The second reduction should occur at approximately midnight and should also be 5 F below the initial reduction. At sunrise elevate the temperature to initial night temperature. Once sufficient light levels have been established in the greenhouse, return the temperature to normal daytime "heat-to" requirement. This process can result in a 10 percent heat savings.

- **Growth regulators** help crop performance. Water is the most effective growth regulator, but in the fast crop system is **not** used as a growth regulator.

Cycocel (chloromequat) is accepted as the most reliable growth regulator. Apply early on in the crop cycle to ensure uniform internode elongation. Make initial application as soon as the crop is established, 14 to 17 days after transplanting.

Mix the chemical to provide a concentration of 750 ppm active ingredient (1:160) and apply at the rate of 1 pint per 100 square feet. Make a second application 14 days later. Make subsequent applications at two-week intervals, allowing at least two weeks between the last application and sale date.

Bonzi (palcobutrazol) also provides effective height control. Apply dilute concentrations of 2 to 5 ppm a.i. at the rate of 1 pint per 100 square feet, 21 to 24 days after transplanting. Due to the highly active nature of this chemical, individual experimentation is strongly recommended to determine the specifics for successful application.

Inverted day/night temperatures (cool days/warm nights) are also highly effective in controlling crop height. It is imperative that the maximum differential not exceed 7 F when employing this technology.

Many of the latest variety introductions require little, if any, growth regulators, due to their genetic backgrounds.

- **Supplemental CO_2** results in faster plant growth and only costs fractions of a penny. Inject CO_2 to provide a concentration of 750 to 1,000 ppm at crop level during peak light periods.
- **Supplemental HID lighting**, at moderate levels of 200 footcandles, can induce earlier flower response and increase flower production.
- **Pest problems** are few. Watch out for caterpillars, aphids, whiteflies (especially on white and purples), thrips (especially on ivy leaf types) and spider mites (also on ivy leaf types).
- **The major disease** problem with geraniums is Botrytis. Proper environmental control that reduces relative humidity and eliminates condensation in and on the leaf canopy is critical.
- **A gradual toning** of the plant tissue successfully completes the process and results in a product that stands up better to the rigors of the retail environment. This can be accomplished by reducing the frequency of liquid fertilizer and water applications 7 to 10 days prior to the anticipated sale date.

"

Joe O'Donovan is marketing manager, Vegetative Geraniums, Goldsmith Seeds, Gilroy, California.

Pelargonium

Growing with Seeley

March 1984

by John G. Seeley

Here we are in the month of March. How about some thoughts on geraniums?

Geraniums from cuttings

Is it too late to propagate and plant? No! It's OK if you do it correctly and can have well-rooted cuttings by April 1. Remember way back in 1946 when the late Kenneth Post of Cornell publicized salable geraniums six weeks after planting well-rooted cuttings in 4-inch pots? Actually, with an April 1 planting, 50 percent of the plants were in flower and salable by May 15, and another 40 percent by Memorial Day. But remember, that was before the day of culture-indexed and culture-virus-indexed (CVI) stock. Cuttings from these plants grow faster than cuttings from ordinary stock.

What must you do for "fast cropping?" Select cultivars that have worked well in research: Ricard, Sincerity, Mme. Landry, Irene, Salmon Irene, Blaze, Pink Camellia, Didden's Improved Picardy, Cherry Blossom, Mrs. Parker, and Velma Cox. Try others if you have favorites; be sure to keep records on time of bloom.

Pot 3- to 4-inch cuttings from indexed stock (preferably CVI) in a well-drained, loose soil mix or peat-lite mix. Keep them warm with minimum of 60 to 62 F at plant height (night) and 70 to 72 F during the day. Be sure the root medium is continuously moist; moisture stress slows down growth.

Furnish a steady supply of nutrients by using a 15-15-15 fertilizer solution at every watering. The solutions should have a concentration of 200 parts per million (ppm) of nitrogen (N). This is easily done by dissolving 18 ounces of fertilizer in one gallon of water and running it through a 1:100 proportioner. This will produce about 200 ppm N at the end of the hose.

If you incorporated adequate superphosphate in the original root medium, then you could use a low phosphorus fertilizer such as 15-0-15 at 18 ounces per gallon of stock solution or 20-5-30 at 13 ounces per gallon.

Give full light for maximum photosynthesis. Do not pinch because that delays flowering. With adequate spacing, geraniums branch well without a pinch.

Geraniums from seed

Seed should have been planted in late January or early February for mid- to late May flowering in northern states. Right now you may be wondering whether your plants are coming along fast enough.

The factors affecting growth of geraniums from cuttings have similar effects on seed geraniums. Temperature is a very important factor.

At Michigan State University, Royal Heins grew Sprinter Scarlet at 62 F night temperature until the inflorescence was first visible when the leaves were separated. Then he switched plants to night temperatures varying from

Pelargonium

50 to 72 F with 5 to 7 F warmer air temperature in the daytime. Vents were opened when day temperatures became 10 to 12 F warmer than NT.

In Experiment 1, started in late March, every drop in NT resulted in a 1-day delay in flowering. In Experiment 2, started in mid April, the delay was ½ day per drop. For example, Royal's data showed that in the late March/early April period it took 38 days from visible bud to flower with 55 F NT; at 60 F it took only 33 days, and at 65 F, only 27 days.

In Experiment 2, where NT did not always drop to the desired levels, the increase in growth rate was less. At 55 F NT it took 27 days; at 60 F the span was 24, and at 65 F it was 21 days from visible bud to flowering. Thus you can easily speed up development by raising night temperature.

Preventing petal shatter

In 1981, William Carlson of Michigan State University reported a 20-day control of petal shattering by applying a spray of a 50 ppm sodium thiosulfate (STS) solution just as flowers begin to open.

Making the STS solution: Terry L. Humfield of the Ohio State University Agricultural Technical Institute described the technique in the December, 1983, *Ohio Florists Association Bulletin 650*. He reported that petal abscission could be delayed up to 30 days.

Step 1: Weigh 20 grams (¾ ounce) of silver nitrate and dissolve it in one pint of distilled water.

Step 2: Weigh 120 grams (4½ ounces) of sodium thiosulfate prismatic and dissolve it in a separate pint of distilled water. If anhydrous sodium thiosulfate is used, weigh out only 2¾ ounces.

Step 3: Then pour the silver nitrate solution slowly into the sodium thiosulfate solution while vigorously stirring to mix the two solutions. This gives you one quart of concentrated stock solution.

Step 4: To make the dilute spray solution, add two teaspoons (1½ fluid ounces) of the stock solution to one gallon of water.

Step 5: Spray the plants at the rate of 10 milliliters (½ of a fluid ounce) per plant just as the florets start to show color. That is not much solution. The one gallon of spray solution should be enough to spray about 375 plants. That would be about one gallon to 100 square feet of bench space if 4-inch pots are on a 6-inch by 6-inch final spacing.

Apparently rates can vary a bit. John C. Peterson of Ohio State University reported that he used ¹/₁₀ ounce of the stock solution per gallon of water and applied this spray solution at the rate of one gallon per 200 square feet of bench space when the first cluster showed color in the tips of the buds. Three weeks later, plants displayed a minimal amount of shattering when plants were shaken. So you may want to experiment, using the two rates mentioned, to observe results under your own conditions.

The January 1982 *Grower Talks*, mentioned Will Carlson's recommendation that growers use a mask or respirator when applying the spray. For a photographic illustration of the effectiveness of the STS technique, refer back to page 23 of your March 1982 *Grower Talks*.

Now for the "sticky question." Is it legal to use? If one concludes that it falls in the category of pesticides and growth regulators, then it should be labeled for this use; it is not. So technically, I cannot recommend or suggest the use of the STS solution for control of petal shattering. But this reminds me of the situation with Cycocel and A-Rest which research has shown to be effective in causing earlier flowering. But neither of these growth regulators is registered by the EPA for use on geraniums. Hopefully, plant breeders will bring out some non-shattering hybrid seed geraniums just as they developed more compact growth habits and a wider range of colors.

A final thought about geraniums

Do you want more information about geraniums? The 410-page manual, *GERANIUMS III*, is a bargain; write to Pennsylvania Flower Growers, 102 Tyson Building, University Park, Pennsylvania 16802.

"

John Seeley is retired form Cornell University.

Perennials

Perennial culture 101:
A how-to handbook of the top six

March 1990

by Miriam Levy

Perennials are definitely here to stay. Here's a cultural guide—including planting instructions and overwintering, soil nutrition and light requirements—for the top six staples: daylily, hosta, astilbe, rudbeckia, coreopsis and sedum.

Perennials are the crop of the '80s. The last 5 years have seen an explosion of sales and of growers entering the perennial market. Unfortunately, as quickly as growers have jumped on the perennial bandwagon, many have jumped off.

One reason for this is a lack of written cultural requirements necessary for producing quality crops. This has changed as university and industry research have increased, and now the production of the top six perennials can be easy and profitable for all growers.

Basic perennial culture

Perennials are versatile and flexible in their growing requirements, adapting to a wide range of environmental conditions. As with all crops, a well-drained soil is a must. Most perennials can be planted in either fall or spring; this decision lies with the grower.

An operation that produces bedding plants may not have labor available to spring plant. Fall planting can occur after spacing poinsettias and shipping

Perennials

garden mums, when labor is plentiful. One benefit to fall planting is cool autumn weather, which is ideal for root development. Then, when warm spring days arrive, all plant energy goes to shoot and flower production.

All perennials require some protection from winter's harshness. The key to successful overwintering is establishing a healthy root system before soil freezes and removing winter protection as growth begins in spring. All varieties must be hardened off before winter; accomplish this by eliminating fertilization, reducing irrigation and exposing plants to a hard freeze. Another factor to be considered is the hardiness rating; not all perennials survive all winter temperatures. As soon as new growth begins in spring, winter protection must be removed and a thorough irrigation applied. To protect newly emerging shoots, it's recommended that any winter covering be removed on a cloudy day.

Daylilies are No. 1

Hemerocallis (daylilies) are at the top of the perennial popularity list. Daylilies are available as field grown rootstock that are commonly dug and divided in late summer. This makes fall the ideal time to pot them. Depth of planting should follow the field soil line; planting too deeply results in crown rot.

Daylilies have thick fleshy roots that store their food, so a low rate of general purpose slow release fertilizer incorporated into the soil or topdressed is sufficient. They tolerate wet soil, as well as reduced moisture and reflected heat and grow well in either full sun or a partially shaded location. Disease is non-existent, but thrips have been known to feed on their leaves.

Overwintering daylilies is a breeze. Minimal winter protection is necessary; a 2-inch layer of leaves or mulch will suffice. They have no special requirements once growth begins in spring.

Day lily

Perennials

Rudbeckia

For shady spots, try hosta and astilbe

Hosta and astilbe are the two top selling shade perennials. Hostas are easy to grow and come in a wide variety of heights, leaf sizes and colors. The varieties most requested by landscapers are *H. albo-marginata* and *H. sieboldiana*.

Availability is commonly by field stock, but there are varieties being produced from tissue culture. Hosta are dug, divided and transplanted from spring until fall. They grow well in almost any soil, but prefer slightly acidic conditions, low fertility and high amounts of organic matter for optimum growth. Planting depth should follow the field soil line.

Hosta do not like to be extremely wet or dry; keep soil damp to the touch. Most hosta won't tolerate full sun; filtered or partial shade is preferred. Insects and diseases ignore them, but an occasional slug may feed on leaves. Winter protection is identical to daylilies.

Astilbe are also propagated from field divisions and come in a broad spectrum of whites, pinks and reds. Rheinland, Peach Blossom, Bridal Veil and Red Sentinel are the more common varieties.

Plant astilbes either in fall or early spring at a depth slightly deeper than the original soil line. They require a cool, shady location in a soil that is very well drained with high amounts of organic matter and fertilizer.

Irrigation is the key to producing top quality astilbe. If the soil dries out once, the leaves will burn and bud abortion is likely. It's almost impossible to overfeed or overwater, and spring growth requires water and food. Astilbes need good winter protection—an unheated house or a 3- to 4-inch layer of leaves or mulch.

Perennials

Rudbeckia cannot overwinter wet

Landscapers love rudbeckia for the deep yellow flowers they produce from midsummer to frost. Goldsturm is the most common variety grown. They are produced either from seed or division. Sow seed in mid July for transplanting in early fall. Purchase rootstock either in fall or spring.

Rudbeckia are very sensitive to crownrot and shouldn't be planted deeply. Always follow the depth of the original soil line. Apply a preventive fungicidal drench immediately after potting. Full sun is preferred, but light shade is tolerated, although it can reduce bloom number. Rudbeckia are heavy feeders and require a soil with high amounts of organic matter.

An overwintering site that provides protection from temperature fluctuations and the wind is necessary. Rudbeckia don't tolerate wet soil during overwintering, and it is imperative that they aren't placed in a low area of a structure where puddling could be a problem during winter. Spring growing requirements are minimal—water and fertilizer produce a top quality rudbeckia crop.

Coreopsis blooms all summer

Coreopsis is another easy-to-grow perennial staple. There are two types: *C. verticillata*, which has thin needle-like leaves and is produced by cuttings or division, and *C. lanceolata*, which is produced from seed.

The most commonly grown verticillatas are Moonbeam and Golden Showers. Cuttings and/or rootstock can be potted in either fall or spring. These varieties have shallow, fibrous roots. Plant rootstock on the soil surface with a handful of soil placed on top of the root to prevent the crown from drying out. They are medium feeders—fertilizer can either be topdressed or added to the soil.

In the spring, after new growth has reached a height of 8 to 10 inches, prune verticillatas back to 4 to 6 inches to increase branching and stagger the bloom time. This will provide flowers from early summer well into fall and can be executed successfully through the fourth of July.

The newest of the lanceolate varieties is Early Sunrise. Seed can be sown September through January with Mother's Day being the natural bloom time. Early sunrise is a photoperiodic long day plant and responds to a 13-hour day. Plants grown from seed sown in April bloom the fourth of July and continue blooming throughout the summer.

Both Coreopsis require a protected site for overwintering success. Minimal care is necessary for production of top quality plants in spring.

Ground covers or tall, sedum offers color

Sedum are the last of the top six. They are available in a wide range of heights, colors and growth habits and are produced from cuttings or field divisions.

Ground cover sedums—Blue Spruce, Kamschaticum and Dragon's Blood—are fall planted. These sedum have fibrous roots, so plant on the surface, with care that top growth doesn't touch the soil surface. Once a root system has developed, soft pinch plants in fall to promote branching in spring.

The fall-blooming, tall sedum spectabiles—Autumn Joy, Indian Chief and Stardust—can be planted either in spring or fall. Pinching stems back in spring promotes basal branching. *Sedum spectabile* have large, heavy flowers and may require staking if too many stems are produced.

Both types have thick, fleshy leaves that store water. Sedums thrive in bright, sunny locations and tolerate a dry soil, but won't grow well in a wet environment. Fertilizer requirements are low, and minimal overwintering protection is sufficient.

Perennials

There is no doubt that the popularity of perennials will continue in the 1990s. As perennial research continues to increase, all growers will be able to produce top quality plants to satisfy market demands.

"

Miriam Levy is with Sluis and Groot, Capitola, California.

Culture notes

November 1988

by Jim Nau

Perennial germination
Aquilegia (Columbine), Asclepias (Butterfly Weed), Centranthus (Red Valerian or Kentranthus), Delphinium, Digitalis (Foxglove), Helianthemum (Rockrose), Hibiscus (*Hibiscus moscheutos*)

Seed germination is still the leading form of propagation for many kinds of plants today, but throughout the maze of different classes and varieties of crops available, there are none more difficult to germinate than perennials and tropical foliage seed.

Perennials are difficult to germinate since their needs can differ from genus to genus, and in certain cases, from variety to variety. Techniques include scarification, alternating day and night temperatures, stratification, chilling and frost germination. While other methods exist, these techniques represent the major pre-germination methods that are commonly used today.

- **Scarification:** Nicking the seed coat to allow for water absorption is called "scarification." This can be done using either a finger nail file or sand paper for rubbing the outer coat of the seed. Use extreme care not to damage the embryo inside. Scarification can also be done chemically with a diluted solution of hydrosulfuric acid (H_2SO_4), which breaks down the seed coat. If the food reserves or embryo are damaged, however, germination rates will decrease substantially.
- **Stratification:** This is the method of placing the seed between layers of moistened paper towels that are in turn sandwiched between moistened sand layers. The layered seeds are placed in a refrigerator in a temperature range of 32 to 35 F for several weeks to several months.
- **Chilling:** This is the method of taking the seed packet and placing it in a temperature range between below freezing to 40 F, depending on the variety.
- **Frost germination:** A frost-germinated seed is a variety that requires the alternate freezing and thawing of the midwinter to early spring weather that the Chicago latitude offers. Seed is sown onto a germination media in a wooden flat and covered with a layer of plastic. The seed is allowed to swell for several

Perennials

days on a germination bench at a temperature from 50 to 60 F. The flat is then placed in a cold frame or other unheated area away from direct sunlight.

The major area of confusion with frost germination is the length of time required to break the dormancy cycle the seed is in prior to germinating. To get the greatest percentage of germinated seeds using this method, you may want to sow seed in the fall and place outside by mid- to late December. Cover the flat with snow, if possible. The dormancy period referred to earlier is broken after a period of temperatures alternating between 32 and 41 F. Finally, bring the flat back into the greenhouse during March, raising the temperatures gradually. Sharp increases in temperature will burst the seed and kill it.

Standard germination techniques

Now that you have the dramatic methods of germinating perennial seed, you'll want to know that there are a wide variety of perennials that will germinate quite well using the standard techniques that are used on annuals and bedding plants.

The following provides specific information on how to germinate a number of perennial seeded varieties. Note that a number of the varieties listed don't incorporate the methods provided above. If germination rates of the perenials you are growing from seed are not as high as you would like, contact someone you know who has knowledge of perennials and can direct you to one of the methods above that may help increase your germination rates.

- **Aquilegia** (Columbine) has irregular germination rates. Keep in mind that most seed sales are in mixes as opposed to separate colors. Chill seed for two to three weeks at 40 F, especially if you have seed left over from last year that you are now using.

In experiments that we have conducted, we have found that by using a fungicide at half the recommended rate watered into the seed flat prior to or at the time of sowing, germination rates have increased by 10 to 15 percent. This technique was not used in combination with any other. This information is provided to show current research and is not a recommendation for use.

Use a soil temperature of 70 to 75 F, leave the seed exposed on the soil surface, and seed will germinate over a period of 21 to 35 days.

- **Asclepias** (Butterfly Weed): I often hear that every grower has a nemesis when it comes to producing one crop or another. For me it's seed-grown dahlias—I never can get them to flower they way I would like. In perennials, several growers indicate that they can never get the seed to germinate at a rate better than 30 to 40 percent on asclepias. It's one crop I haven't had a problem germinating, although seed kept from year to year does germinate better if stratified the second year.

The period after germination, however, is critical for asclepias. Too many times asclepias seedlings will die in the flat due to a restricted cell pack. Asclepias has a large taproot that will die if not allowed to develop properly. For germination, use a soil temperature of 72 F, leaving seed exposed to light upon sowing. Germination occurs between 21 and 28 days.

- **Centranthus** (Red Valerian or Kentranthus): Centranthus is not as common as the previously two noted varieties of perennials but it is definitely worth trying. In our tests, centranthus germinated well using standard bedding plant procedures. Use a germination temperature of 60 to 65 F and lightly cover the seed. Germination occurs in 14 to 21 days, and plants will flower in as soon as 16 to 18 weeks.

- **Delphinium:** Like its cousin, the larkspur, delphinium is not well known for its high germination rates. Approximately 60 percent of the world production of delphinium seed occurs in fields within the United States and is harvested in

Perennials

late summer to early fall. The primary trouble with delphinium is that the seed goes into dormancy within several months after harvesting.

The most noted method for breaking dormancy is to chill seed at 35 F for at least one week and up to four weeks prior to sowing. Some work has also been done on breaking the dormancy by using gibberellic acid. This is provided to show research and is not a recommendation for use.

It has also been shown that varieties like the Pacific Giants and Centuries germinate best when given an alternate day temperature of 80 F and night temperature of 70 F. Other varieties, like Belladonna and Bellamosum, germinate using a constant temperature of 65 to 75 F. Upon sowing, cover the seed lightly for germination, which will occur in 10 to 18 days.

- **Digitalis** (Foxglove): This is one of those crops you should have little trouble germinating. In our tests—regardless of whether you covered the seed or not—using a constant soil temperature between 72 and 75 F gave a germination rate of no less than 80 percent. Seed will germinate in five to 10 days.
- **Helianthemum** (Rockrose) will normally germinate in a range of 15 to 30 percent if the seed is not given a pretreatment. Helianthemum has a high percentage of "hard" seed that requires a scarification process prior to sowing. Though tests are not complete yet, some initial results indicate that treating Helianthemum as a frost germinater is useful in increasing germination percentages. For germination, use a soil temperature of 70 to 75 degrees F, either covering the seed or leaving it exposed to light. Germination occurs in 15 to 25 days.
- **Hibiscus** (*Hibiscus moscheutos*): This is the perennial hibiscus that we grow here in our gardens in West Chicago, Illinois. Hibiscus from seed is not well known for its high germination rates.

In one simple test, we placed seed into warm water and let the container stand overnight prior to sowing. We noticed that a number of seeds had continued to float on the surface instead of sinking to the bottom. If the seed is still alive, nicking the seed coat and placing the seed back into the water increases the chances of germination.

In our tests, the seeds that didn't sink were sown onto the media along with the seeds that did sink. Germination occured on both lots, but germination occured more readily and with a stronger stand for seeds that sank, rather than for those that stayed afloat.

Another test we tried was to chill the seed at 32 F for a week prior to sowing. This method also decreased germination time. Compared to seed that was sown with no pretreatments, chilled seed had a germination percentage of 67 percent 13 days after sowing compared to 30 percent for seed with no pretreatments. Use a germination temperature of 70 to 80 F and cover the seed. Germination occurs in seven to 10 days.

"

Jim Nau is trials and product development manager for seed at Ball Seed Co., West Chicago, Illinois.

Perennial Production

June 1988

by Miriam Levy

A grower's notebook for summer

The days of June, July and August traditionally are quiet times for growers. Bedding plant season has peaked, poinsettia propagation is just beginning and for many, summer vacations are the main topic of thought. For the perennial grower however, the days of summer are filled with more than sales of summer blooming plants. Propagating material for fall and spring sales, as well as growing on material for summer and fall sales, keep perennial growers busy all summer long.

June propagation

At Martin Viette Nurseries, East Norwich, New York, perennials are grown year round. With the peak of spring sales behind us, June is the month to begin preparing our fall blooming perennials for late summer and fall sales. Many fall blooming varieties have not yet budded and June is the optimum time to take cuttings for the following fall. Taking cuttings also acts as a pinch and provides a fuller plant with more blooms that year. Our fall sales season begins in mid-August—eight to 10 weeks from taking a cutting is sufficient time for the plants to fill out and set bud. These varieties include Sedum, Cimicifuga, Eupatorium, and Montauk Daisy.

Cuttings are 2 to 3 inches in length and stuck in a mixture of peat and sand. Rooting hormones reduce rooting time on perennials and are used on all propagated varieties. With the humid, hot days of Long Island summer approaching, we have found it unnecessary to use any heat during rooting, the challenge is to keep cuttings cool. Flats of cuttings are placed under either mist, the shade of a tree or in a dark cool corner of a greenhouse and covered with moist paper if a mist system is unavailable.

June days are also spent deadheading any leftover spring blooming perennials. Some rebloom in the fall such as Tradescantia, Ajuga, Anthemis, or Coreopsis; but most produce vegetative growth from which cuttings can be taken later in the summer.

Division of spring blooming perennials is also done in June. After flowering, we allow the flowers to die back naturally before dividing. Siberian, German and Japanese Iris are all divided in June. Divisions are made leaving three to five eyes. Divisions are dipped in a fungicide before planting. Warm season ornamental grasses are also divided in early June. Since these grasses have active root growth during the warm days of summer, mid-June is the latest that division should occur. Warm season grasses include Miscanthus, Pennisetum and Panicum. Divisions should be planted with the new shoots directly below the soil level.

Perennials

June growing-on

In addition to June propagation, summer blooming perennials are actively growing. Most of the early spring blooming varieties are finished for the year and the heat tolerant plants of summer take over: Rudbeckia, Coreopsis, Veronica, Achillea, Lythrum, Geranium, Campanula, Salvia, Phlox and Hosta to name a few. Soil moisture conditions must be closely monitored, since excessive drying may cause marginal leaf burn. We feed weekly with a 200 ppm complete fertilizer throughout the summer. By this time we have reached the seventh month of our eight to nine month slow-release fertilizer, and we have found all varieties benefit from a weekly feed.

Our high humidity conditions of summer provide the ideal environment for disease and insects and a weekly spray program is strictly adhered to. We use a combination of an insecticide and fungicide and rotate the chemicals biweekly. Leaf spot and mildew-prone varieties such as phlox and aster are grown on drip tubes, which significantly reduces the incidence of powdery mildew.

Many of the plants that were deadheaded or pinched in June are ready to have cuttings taken in July. Budded cuttings will not root; any buds that may be present must be softly rolled out for rooting to occur. Cuttings of silver foliaged varieties such as Artemesia and Stachys tend to hold water and rot easily if they're kept too wet.

We find that most perennial cuttings wilt severely the first few days after sticking, but seem to come back without apparent damage. Unlike indoor plant varieties, they seem to have a higher tolerance for less than optimal conditions during rooting, and almost always return to a turgid condition.

Sowing seed in July

Many early blooming rock garden varieties are propagated from seed. We sow into 128-cell plug trays in early July with a vacuum seeder. The larger-sized plug allows us to hold the seedlings through the summer until fall planting without reduction in plant quality or vigor. If limited growing or overwintering space is a factor, seedlings can be overwintered in the plug flats and transplanted in the spring.

We transplant seedlings into final containers beginning Labor Day. This allows ample time for root development, but the cool fall weather limits excessive top growth. If too much top growth does occur, it becomes an ideal location for disease development when the leaves die back in the winter. All of our perennials grown from seed bloom the following spring. Perennials that we seed ourselves include Aquilegia, Campanula, Coreopsis, Shasta Daisy, Lupines and Digitalis. All except Aquilegia are sown singly, with one seedling transplanted into the final container. Aquilegia is triple-sown and three plugs are transplanted. This ensures a full pot in early spring when they bloom.

Varieties propagated in July for sale the following spring
Achillea
Ajuga
Amsonia
Coreopsis
Helenium
Heliopsis
Lysimachia
Lythrum
Monarda
Nepeta
Phlox (tall varieties)
Plumbago
Salvia
Sedum
Stachys
Veronica

Dividing field crops

In late July and August field crops are dug, divided and planted. Martin Viette Nurseries has 5 acres of daylily fields and 4 acres of Hosta stock. We have over 300 varieties of daylilies of which at least 100 varieties are dug each year for sales the following spring. For every clump dug and divided for potting

Perennials

into containers, an equal number are divided and replanted into the ground. Therefore, fresh stock is constantly being grown.

A minimum of three years in the ground is required before a plant is dug for potting. Weeds do exist and in addition to our fall through spring herbicide program, all divisions are cleaned with a high pressure hose before potting.

This season's growth is cut back as close to the root as possible and any new growth that sprouts in the fall is allowed to die back naturally to increase the carbohydrate levels in the roots, which will be utilized by the plant the following spring.

Hosta divisions are treated identically to the daylilies. Cool season grasses are divided in August. These grasses have their active root growth in the fall and division should occur before September 1. Cool season grasses include Festuca, Deschampia and Calamagrostis.

August, buying in plugs

Due to the limited space in our seed chambers, we have begun buying perennial plugs. Many of the annual plug specialists have entered the perennial plug market. We purchase these in late August and pot them up at the same time as our seedlings. Perennial plugs are generally in trays of 400 and we have found that three plugs in a pot produce a top quality plant next spring. Plugs that we buy include Arabis, Aubretia, Cerastium, Aurinia, Pyrethrum, Leontopodium and Oriental Poppy. Every perennial is drenched with an all-purpose fungicide after potting.

By late August our work has paid off and full blooming pots of fall perennials are ready for sale. Japanese anemone, Kirengeshoma, Aconitum and Cimicifuga sell before they bloom. We have seen a definite increase in consumer demand not only for fall blooming material, but for vegetative spring and summer blooming varieties that can be planted in the fall. This is where our summer fertilization program pays off. Plants that were pruned, used for propagation and fed all summer are now lush, green and ready for sale.

The majority of our perennials are grown from rootstock planted in October. With the hectic days of fall planting just around the corner, September is spent in preparation. Equipment is serviced, labels are printed, soil is mixed and stockpiled. We use a mix of peat, sand and perlite with an eight- to nine-month slow release fertilizer containing major and minor elements. Overwintering structures are repaired and their floors regraded since proper drainage is crucial for successful over-wintering. Everything possible is done in advance so when the roots arrive, all we do is unpack and plant, and the year's cycle starts new.

Summer vacations are nice, but for perennial growers there are many ways to make the lazy days of summer productive and profitable.

"

Miriam Levy, formerly with Martin Viette Nurseries, East Norwich, New York, is with Sluis and Groot, Capitola, California.

Poinsettia, see Euphorbia

Primula, Primroses

Culture notes

September 1988

by Teresa Aimone

Primula
Family: Primulaceae
Genus, species: *Primula obconica*

In order to satisfy the grower's constant search for exciting and different pot plant items, seed production companies work hard to find plants that fill both the production schedule and the marketing plan niches.

Many growers have found such an item in blooming F_1 *Primula obconica*. This member of the primula family offers several advantages:
* Greater heat tolerance than other types of primula.
* Large, long-lasting blooms that hold up well in the retail shop and the consumer's home.
* Exciting colors not found in other primulas.
* A compact and mounded plant habit, adaptable to 4-inch, 5-inch or 6-inch containers.
* Good potential for year-round pot plant production.

F_1 *Primula obconica* plants have hairy leaves with slightly ruffled edges. Multiple stems arising from the center of the plant bear large clusters of flowers. Hybrid *P. obconica* is adaptable for 4-inch sales, but provide an even more spectacular show when finished in 5-inch or 6-inch pots.

Production of this interesting pot plant is slightly similar to production of *Primula acaulis*.

Production
Germination: Depending on variety, there are approximately 5,000 to 6,000 seeds per gram and 104,000 to 200,000 seeds per ounce of *Primula obconica*. Seed should be sown in a sterile, porous, peat-lite mixture. Although seed require light to germinate, moisture is critical. Covering the seed with a thin layer of very coarse vermiculite will allow light to pass through to the seed and help retain necessary moisture levels. Allowing the seed to dry out for even 30 minutes will dramatically reduce germination levels. Misting or fogging seed during germination is recommended to obtain vital moisture and humidity levels.

P. obconica seed require a germination temperature of 68 to 70 F. Seed germinate in about 10 to 20 days. Growers may apply a preventative fungicide drench to the medium prior to sowing and at transplanting. Check fungicide labels for proper rates.

Primula

Seed can be sown in rows in open seed flats, or produced in plugs. Large plug cell sizes, 288 and larger, are recommended, once again due to the need to maintain consistent moisture around the seed.

Year-round production is possible. Main sowings are usually in July and August for February to March blooming. These times provide growers with the optimal conditions for *P. obconica* production. Careful attention to temperature and light allows for efficient production during less than optimal times of the year. Depending on pot size, temperature and time of year, crop time ranges from five to seven months.

Transplanting: Approximately three to five weeks after germination, transplant plugs or seedlings. Transplanting can be done directly into the final container, or into an intermediate size to be stepped-up later. Use a peat-lite media with a pH of 5.5 to 6.0.

For 4-inch production, transplant one plant directly as the final planting. For 5-inch and 6-inch production, growers may transplant two plants per 4-inch, allowing plants to establish for approximately four to six weeks, then transplant into the finishing pot. This is not vital for improved plant appearance, but it does save on bench space. Growers can, of course, transplant two plants directly into the larger final container, saving an extra step.

Temperature: In the young stages, grow plants at 65 to 68 F. About one and a half months prior to finished plants, reduce temperatures to 59 to 65 F to produce plants of the best quality.

Light: *P. obconica* do not prefer direct sunlight. Provide light shading during young stages and sunshine. Shading will also keep temperatures in more desirable ranges.

Watering: Keep plants constantly moist during the production stages. Allowing plants to dry out could result in marginal leaf burn.

Fertilizing: Provide plants with a balanced fertilizer application once a week. *P. obconica* have low to moderate fertilizer requirements, and can respond with more vegetative growth than desired when fertilized heavily. Avoid ammonium nitrate forms of nitrogen, especially during winter and with cool temperatures.

One special note: Some growers and greenhouse workers have experienced a skin rash or irritation when handling *P. obconica*. This should not be regarded as a deterrant to production of this exciting pot plant. Provide workers who are handling this product with latex gloves. They should also wear long sleeves if possible. Thoroughly wash or discard gloves after each use. If irritation or swelling occurs, consult a physician.

Varietal selection
Always use F_1 varieties as they produce the best plant habit and flowering, and the shortest production times. Currently available is the Juno series, available in the following colors: Juno Orange—deep appleblossom orange; Juno Blue—bright mid-blue; Juno Rose—bright carmine-rose; Juno Red and White—flowers open to white, then turn from light rose to dark rose as they mature; Juno White—clear white; and Juno Formula Mixture—a blend of the above colors.

99

Teresa Aimone, former editor of GrowerTalks *magazine, is with Sluis and Groot, Fort Wayne, Indiana.*

Culture Notes

July 1987

by Ron Adams

Primula Acaulis are popular perennial bedding and pot plants started from seed. They are cool growing plants and depending on location, sales of finished bedding plants start as early as November in California and southern states. Other parts of the country grow them cool for pot plants, flowering them February through April. There are several species of primula with four main types forced in the greenhouse. The forced types are Primula Acaulis, Malacoides, Obconica, and Veris (polyanthus).

The Acaulis type is identified by its many flowers born in or near compact foliage on individual flower stems. There are many strains of Acaulis types with a made range of flower color and characteristics. Started from seed, it takes primula between 20 and 22 weeks to flower a $4^{1}/_{2}$-inch pot, depending on variety and forcing temperature, Plants are also available from specialist propagators.

Propagation
Sow seeds in a well-drained, fine textured, peatlite media. In open flats sow in rows approximately 1,000 seeds to a 11-inch by 22-inch flat. Single sow in plug trays. *Do not cover the seed.*

Germination: Soil temperature 60 to 65 F; light, no higher than 1,000 footcandles. HID or fluorescent can be used with 12 to 18 hour days, and should be at least 450 footcandles. Germination will start in seven to 10 days and take up to 28 days to complete at 60 F. After germination is complete, reduce soil temperature when possible to 55 F to harden the seedlings prior to transplanting. During the later stages of germination, feed with 150 ppm nitrogen from either a calcium nitrate or potassium nitrate source. Finally, transplant the seedlings at the two to four true leaf stage.

If growing in plug trays, continue the 60 F soil temperature until the seedlings are well established.

Caution: If the soil temperature is too high, the seeds will not germinate properly, resulting in a low percentage or no germination in an extreme case. If your germinating conditions do not allow you to start seed at the optimum temperature, then you will want to consider getting already started seedlings from a specialist.

Soil moisture: The soil moisture level should be uniform since drying-out will reduce germination. The soil should be moist but not saturated. This can be accomplished by misting, fogging, or covering the seed flats with glass or plastic. If you are covering your flats with glass or plastic make sure to keep them out of direct sunlight to avoid heat buildup under the covering.

When seedlings are ready, transplant into 2¼- to 4½-inch pots, planting one seedling per pot.

Temperature: Maintain 55 to 60 F soil temperatures until the plants are established—roots to the bottom and edges of the pot. Begin feeding and allow the plants to dry between waterings once they are established. When the plants have reached the six to 10 true leaf stage, night temperatures can be lowered to 40 to 50 F for the duration of the crop. Forcing at temperatures above 50 F will produce large leaves and can cause the flower to open in the foliage. It is possible that dropping the night temperature before the six to 10 true leaf stage will interfere with the plants' ability to initiate flower buds.

Soil and fertility: Use a well-drained peat lite mix or 2-1-1 (peat, soil, perlite) soil mix. PH of the peat lite mix should be 5.2 to 5.8 and the soil mix 5.8 to 6.2. Feed at a constant 150 ppm nitrogen level with the nitrogen source either calcium nitrate or potassium nitrate. Avoid the use of ammonia forms of nitrogen which can cause tip burn. Because primula are grown at lower temperatures, it is easier to build up high soluble salts. The soil should be tested monthly to avoid soluble salt buildup. If using a peatlite soil use a balanced fertilizer that includes trace elements.

Light: Grow on with a light level of 2,500 to 3,000 footcandles. Typically full sun in the winter when the days are short unless you are in a high sunlight area where levels consistently exceed 3,000 footcandles.

Disease control: Botrytis is the biggest problem with primula. Maintain good air circulation and treat with fungicides as needed. It has been reported that Exotherm Termil, Chipco 26019, Benlate, and Ornalin work well for Botrytis control on primula.

Insect problems: Primula can be infected by aphids, fungus gnat, spidermites, thrips, and whitefly. Check with your local extension specialist for recommended treatment.

Flower forcing: When buds are visible, raise temperature to 60 to 62 F, which will flush the buds into a more uniform bloom. A high temperature (above 62 F) at finish will produce a tall, weak flower stalk.

Premature flowering: A common problem with primula production is premature flowering, brought on by shock or stress such as getting pot bound too soon, or overwatering and losing the root, or allowing the soil to dry out between waterings.

Varieties

Variety selection will be dependent on market and location needs. Consult your seed or plant source to determine the variety best suited to your growing conditions and market needs.

99

Ron Adams is technical services manager, Ball Seed, West Chicago, Illinois.

Primula acaulis:
Culture, scheduling, cooling

October 1985

by Vic Ball

This colorful and very fragrant late winter pot plant is a big winner all over the world. We hear of 500 kilograms of seed used in Japan (1,200 pounds!) and that they are the #1 flowering pot plant in Japan. And the same huge volume grown across England and Europe. They are very big—especially in France and England. They are hardy and a great favorite in the gardens in such mild winter and cool summer areas as England and our own Northwest. But most of the millions of plants sold are just enjoyed for a week or two as a colorful, small houseplant.

One reason for this huge success is the ease of culture—and very low labor and fuel requirements. Again, they do sell in huge volume, so inevitably prices are very competitive. But, when you grow an acre of them, you don't have to make much per plant to make a profit!

They are a favorite of supermarkets and food stores the world over—both because they are a relatively low-priced item, and because they are very colorful and fragrant; therefore, they are a strong impulse seller. By the way, they do come in a wide range of very bright, cheery spring colors. We've seen them offered in European food chains at around 70¢, U.S. Virtually all the European and Japanese crop is grown in 9 cm to cm pots—about 4 inches. Growers normally figure 40 to 45 plants per square meter.

One more comment on their popularity around the world—European seed firms report them far bigger in seed sales than seed geraniums or even impatiens (as pot plants).

Three primula families

You'll commonly find these three primula families in the commercial seed catalogs: *Primula acaulis*, the Julians and *Primula veris*. Here's how they sort out in the real world.

- **P. acaulis:** Probably three-quarters of all the group of primulas grown are *Primula acaulis*. The Festive series is very popular, the Pageants also widely used and earliest. Ducat, Disco and Sunset are other common varieties of *Primula acaulis*. The culture we've talked about in this story applies to the acaulis variety.
- **Julians** are really an acaulis primula. The difference is that the Julians offer smaller foliage, a smaller flower, but a great deal more of flowers. Culture the same as the regular acaulis mentioned above. Julian Mixture, Cheerleader, Cherriette and Gold Ridage are Julian varieties.
- **Primula veris** is a different series. Here, all flowers occur in a sort of cluster, from one main stem. Otherwisw, they are basically like the other Primula veris—including culture. Popular varieties are Elite Mixture, Pacific Giants Mixture and Jewel Mixture.

Primula

A block of Vens Primula in Denmark on November 5—outdoors! They were scheduled to go under glass that week. Growers: the Commerou Brothers, near Copenhagen.

Schedules

These primulas are, of course, seed-propagated—and basic sowings occur in summer or early fall—for flowering mainly in February-March. They are a special favorite for Valentine's Day flowering. Here are several most used approaches.

- **Northern United States:** Seed is sown early—mid-September for flowering in February and March in 4-inch. One problem: if 90 F temperatures occur, germination will be poor or nil. At the other extreme, they need about a 68 F soil temperature for good germination—no less. One answer to the problem of hot weather during germination is to use plugs—freely available now from specialists in the Northwest where they don't have 90 F weather.
- **Northwestern United States, British Columbia, other cool summer areas:** *Primula veris* are a major fall bedding plant, especially in the Washington, Oregon and Vancouver areas. Millions of plants! These growers normally sow in June or July and offer 4-inch plants for outdoor planting in the fall. In this climate, they tend to flower a bit in the fall and again in early spring—much like pansies in our deep South.
- **Europe:** Kees Sahin, Dutch breeder and seed supplier to whom we are indebted for many of these notes, suggests that main sowings in Europe occur in June and July. "They can be sown from April through early September— which would produce perhaps a late March or early April flowering." (September sowing.) Kees also suggests that if they are sown as early as April, they tend to get leaf spot and also may produce a leafy, inferior-type plant—result of a high temperature. He also suggests that some growers in Europe use artificial growth chambers to maintain cool enough temperatures for good germination. And lastly, he also suggests the alternative of plugs—which are becoming important in our country.

Speaking of plugs, Kees suggest sowing July 1, transplanting plugs to the finished pot September 15 to October 1.

Temperature needs

A common practice in the United States is 68 F soil temperature for germination, then 60 F for several weeks after transplanting—to establish roots—get the plant started growing. Very important! After this 60 F period, they should be grown on at 50 F—until plants have achieved their full size. Then (critically important) they must be dropped to 40 F for a minimum of three to four weeks. This 40 F cooler temperature hastens flowering a great deal—and tends to improve quality. Very Important!

Primula veris always want full sun, but they seem to be able to flower very nicely in our own Midwest even with the limited amount of sunlight available through our winter.

Primulas tend to develop iron deficiency, which results in yellow leaves.

Sales trends on primulas are definitely up, especially in our Northwest. Also, across the country, we see much more of growers buying in 2¼-inches and plugs—produced by specialists. A way around the summer germination problem.

Acaulis are great!

Roland Godard, an English researcher, offers an interesting comment on these primulas. We quote: "In which species under the sun can we find a color range that includes everything from white, blue, to scarlet, not to mention numerous shades of pink, coral, orange and plain colors? Or find flowers enhanced by a dark eye or an edging? Primulas.

"Which species is able to save a lot of that precious energy everybody is taking such great care to conserve, able to thrive at 41 F and can even withstand frost? None other than primulas.

"Which species provides such brilliant, sparkling and dashing colors from fall to spring and even on the dullest days of winter—livening up the home, window boxes, tubs and beds later on? Primulas, of course.

"There's no question which species is today's top winter pot plant crop in Europe. Primulas."

"

Vic Ball is editor in chief of GrowerTalks *magazine.*

Radermachera, China Doll

Culture notes

December 1988

by Ron Adams

China Doll
Family: Bignoniaceae
Genus, species: *Radermachera sinica*

Radermachera is a relatively new introduction to the foliage plant business. Its success for interior plant usage was first recognized in Europe, and its popularity has consistently increased since introduction.

Radermachera is native to Asia and is seed started. Some specialist propagators have provided seedlings and liners to the trade. Radermachera has a rich, glossy appearance with profuse, feathery leaves. It is more compact and faster growing than leea or schefflera, and adapts readily to an indoor environment.

Seed Sowing

Sow seeds on a moist, peat-lite medium, and water in gently to bury the seed slightly. Flats can be covered with clear plastic and kept at a 70 to 75 F soil temperature. Germination should occur in seven to 10 days.

Caution: Radermachera seed can lose its viability quickly. Seed should be sown shortly after arrival or stored in an air-tight container and kept in a cool, dry place.

After germination, grow on to transplant size at a 65 F minimum night temperature. Feed 150 parts per million (ppm) N with a 20-10-20 peat-lite, and keep soil evenly moist. Light levels should be 2,500 to 3,000 footcandles until transplanting. After transplanting, seedlings should be grown at 60 to 65 F

Scheduling				
Pot size	Plugs per pot	Liners per pot	Time to finish	
			Plugs	Liners
4 inch	1	1	6-8 weeks	4 weeks
6 inch	3	1	12-14 weeks	8-10 weeks
8 inch	3	1	18-20 weeks	14-16 weeks
10 inch	3-5	1	24-26 weeks	20-24 weeks
14 inch	3-5	1	38-46 weeks	34-42 weeks

Radermachera

minimum night temperature. Light levels should be in the 2,500 to 3,000 footcandle range for two weeks, followed by a minimum of 2,500 footcandles up to full sun. Plants grown at full sun levels will have to be acclimated at a foot candle level of 500 to 1,000 for four weeks prior to interior usage.

Media and Feed

A well-drained media is best with 25 percent air porosity level. Soil should be kept evenly moist. Drying out causes lower foliage to drop. Feed levels vary from 150 to 250 ppm N, depending on light conditions. A balanced feed is best—20-10-20 peat-lite or 15-16-17 peat-lite, and pH should be 5.8 to 6.2.

Transplanting

Plants should be set at the original soil line. Do not bury the soil ball, but keep flush with the original level. When growing from plugs, figure one plug per 4-inch plant, three per 6-inch plant, and three to five plugs per 10-inch and larger plants. Pinching is not recommended since the plant will only yield one to two breaks. It is best to use multiple seedlings instead of pinching to produce bushy plants.

Growth Regulators

B-Nine is effective for controlling internode length. Concentrations vary from 1,500 to 2,500 ppm depending on location or season. The first application should be made two to three weeks after planting in the final container, with multiple applications repeated every 10 to 14 days depending on the type of growth desired. A final application of B-Nine at 2,500 ppm one to two weeks prior to shipping will hold the shape of the finished plant.

Disease Control

Radermachera is relatively free from diseases. Maintaining good cultural practices is the best disease prevention. Keeping foliage dry at night with good air movement reduces foliar disease problems. Radermachera is sensitive to high ethylene levels, which cause excessive leaf drop.

Insect Control

Radermachera is subject to infestations of aphid, whitefly, mealybug, spider mite and scale. A regular program of insect control should be established and maintained throughout the crop cycle.

"

Ron Adams is technical services manager for Ball Seed in West Chicago, Illinois.

Ranunculus

Culture notes

November 1987

by Lewis Howe

Persian Buttercup
Family: Ranunculaceae
Genus, species: *Ranunculus asiaticus*

Ranunculus, or the Persian Buttercup, native to Southeastern Europe and Southwestern Asia, is used extensively in the summer or grown in a greenhouse during the winter.

It is said that Louis IX brought the first bulbs from the Holy Land into France. The bulbs were brought into England in 1596 where breeding work was performed. By 1830, over 500 varieties existed. In the late 1960s, breeding work produced genetically dwarf pot types from seed.

Today, dwarf ranunculus seed is available. The plants are easier to grow, having a uniform habit and short peduncles. While the growing cycle is longer than other pot items, Ranunculus are economical because they are grown in a cool greenhouse.

Propagation

Bulbs: Production of this crop is simple from tuberous roots. Begin with top grade bulbs size 1½- to 2¼-inch (4 to 6 centimeters) in circumference. Tuberous roots are sprouted in large flats or directly in the final container. The roots can be planted into 4-, 5- or 6-inch pots to save the transplanting step.

One tuberous root per 4-inch pot or two to three tuberous roots per 5- or 6-inch pot produces a quality pot plant. Keep the soil temperature in the flats or pots 65 F at night. Roots begin to sprout in two to three weeks.

If the tuberous roots were sprouted in flats, replant in the final pot when two to three leaves have expanded.

An open, well-drained growing medium is critical to growing ranunculus successfully. Some growers use a 50:50 peat/perlite medium to obtain good drainage. Crop time is 16 to 17 weeks from planting.

Soon, pregerminated tuberous roots will be available to commercial growers. Growing procedure is the same, but crop time is only six to eight weeks to flower. Upon receiving the tuberous roots, plant immediately. Initially, a few varieties will be available.

Seed: Ranunculus seed is available from many seed companies. There are 42,500 ranunculus seed per ounce (1,500 seeds per gram).

Sow seed evenly in rows in germination trays. A medium of 30:70, peat/perlite medium accelerates seed germination and seedling growth. Cover the seed thinly with the peat/perlite medium and water the seed flats thoroughly. Place in a well ventilated area and avoid direct sunlight.

The seed germinates at soil temperatures of 50 to 60 F in 14 to 21 days. Do not exceed 68 F germinating soil temperature. Never allow the medium to dry out until the germination period is complete.

In warm climates sow seed on October 1 and in cold climates sow seed September 1 for greenhouse flowering March through April.

Plugs: Seed germinates in 14 to 21 days at 50 to 60 F soil temperatures. Most Ranunculus plugs are produced in a 1-inch plug to promote a vigorous root system, which is critical to quality crop production. Sow seed into 98- or 128-cell trays. After germination is finished, grow seedlings at 55 to 60 F nights to speed root development. Day temperatures should not exceed 77 F during the growing period in the plug tray. Maintain even moisture levels and apply a liquid fertilizer, such as 15-16-17 at 100 ppm nitrogen beginning at the first true leaf stage. Do not overfertilize as the plants can get too lush.

The production time in a 4-, 5- or 6-inch pot is 19 to 22 weeks, from seed. Growers can buy in plugs and reduce crop time to 10 to 12 weeks.

Growing on

Transplanting: For 4-inch pots, thin the seedlings to 1 inch apart in germination trays. After four to five true leaves develop, transplant into 4-inch pots.

Ranunculus

For 5- and 6-inch pot culture, transplant seedlings at the two true leaf stage into 2¼- or 2½-inch pots. Repot plants into 5- or 6-inch containers when the root system reaches the bottom of the pot.

Prick seedlings carefully, not to damage root systems. Growth is slow for two to three weeks until root systems are established. Space final containers so leaf tips do not touch allowing for maximum growth and less disease.

Lighting: Grow Ranunculus under a short day length, eight to 10 hours of light. Long day lengths of more than 10 hours result in poor growth and delayed flower bud formation.

Temperature: During the growing stage (before visible bud), grow Ranunculus at 55 to 60 F day temperatures and 40 F night temperatures. As flower buds are visible, adjust night temperatures to 45 to 50 F and day temperatures to 60 to 68 F for early-sale flowering pot plants. Higher temperatures at visible bud stage produce long petioles and peduncles and excessive leaf growth. Ranunculus can be produced in a non-heated greenhouse at 32 to 35 F after being placed in the final container, but production time is increased four to six weeks.

Fertilization: Ranunculus require moderate to high nutrition. If grown in a heated greenhouse, topdress with Osmocote 14-14-14 at ¼ teaspoon per 4- or 5-inch pot and ½ teaspoon per 6-inch pot. Apply a complete liquid fertilizer—15-16-17-at 250 ppm nitrogen once or twice a week. If constant liquid feeding is preferred, apply 250 ppm to 300 ppm nitrogen.

If grown in a non-heated greenhouse, use a complete liquid fertilizer only at 200 ppm nitrogen at each watering. The plants require less watering and nutrition when grown cool.

Growth regulators: If grown in southern climates, apply B-Nine at 2,500 ppm as buds first appear in the crown. As the peduncles elongate, lower growing temperatures and let dry somewhat between waterings.

Insects: Aphids, leafminers and spider mites can attack Ranunculus crops. Oxamyl 10G applications three to four weeks after transplanting the tuberous roots or seedlings will control these insects. Orthene 75 SP will control aphids and leafminers.

Cultivars

Bloomingdale: The first, true F_1 hybrid from seed, Bloomingdale is excellent for 4-, 5- or 6-inch pot sales. Extra dwarf plant habit of 8 to 10 inches (20 to 25 centimeters) with small leaves. Flowers are fully double, 3½ inches (9 centimeters) in diameter borne on short, sturdy stems. Extremely hardy and can withstand temperatures as low as 23 F. Produces flowering plants in 19 to 20 weeks from sowing, perfect for late winter/early spring pot plant sales. Available in separate colors of Pink, Rose, Red, Yellow and White.

Pot Dwarf Mixture: Early flowering type with large flowers 3½ inches (9 centimeters) in diameter. The mixture includes scarlet, red, rose, pink, orange, salmon, yellow, white and intermediate colors. Plants stand 10 inches (25 centimeters) tall.

Bulb varieties: Most bulb varieties are listed by color, not by a specific name. Also, consult supplier catalogs for cut or pot types.

99

Lewis Howe is a research assistant with Park Seed, Greenwood, South Carolina.

Rhipsalidopsis, Easter cactus

A grower's guide to commercial production of Easter cactus

November 1989

by Thomas Boyle and Dennis Stimart

Looking for a new pot plant to boost year-round sales? Try Easter cactus, a crop suitable for 4- or 6-inch pots or hanging baskets. Experts have put together culture tips from propagation to finishing for you.

Easter cactus (*Rhipsalidopsis* cultivars) is a relatively unknown crop in North America, but has been grown in northern Europe for many years. Plants are comprised of leafless, flattened stem segments (or phylloclades) that are similar to the common Holiday cactus (*Schlumbergera* cultivars). The star-shaped flowers range from pink, lavender and fuchsia shades to scarlet and crimson.

Cultivars. Only a few cultivars are used for commercial production. Crimson Giant, also known as Scarlet O'Hara or Giant, has 1½- to 2-inch, scarlet-red flowers and is the most popular cultivar in North America.

European breeders have developed most commercial cultivars. For example, Red Pride (bright red), Purple Pride (deep fuchsia), Evita (purple) and Rood (red), bred by J. de Vries Potplantencultures of Aalsmeer, Holland, are popular in Europe. These cultivars are being tested at the University of Massachusetts and University of Wisconsin.

Propagation. Propagate plants by rooting mature phylloclades in cellpacks or small pots. Stick two or three single phylloclades in each cell or a 2¼-inch pot. Phylloclades can also be direct-stuck in 3- to 4½-inch pots. Use a well-drained, pathogen-free medium for propagation and maintain medium temperature of 75 to 80 F during rooting.

Phylloclades root equally well with intermittent mist, high humidity or periodic handwatering, as long as the rooting medium remains warm and well-aerated. Application of a broad-spectrum fungicide after sticking will reduce incidence of basal rot.

Fertilization and growing media. Start fertilization once roots reach the sides of the propagation container. Fertilize plants with 150 to 200 parts per million nitrogen once or twice a week, using a complete (N-P-K) fertilizer. Leach occasionally with tap water to reduce soluble salts accumulation.

A well-drained growing medium is essential for Easter cactus production. Free drainage is important for reducing incidence of root and basal rot. Many growers prefer to use a soilless, peat-based medium.

Transplanting and spacing. Plants require eight to 16 weeks from propagation until transplanting, depending on the time of year when propagation is initiated (longer in winter and shorter in summer). Each plant should have two or more phylloclades at transplanting. Use one cell or small pot (containing two or three individual phylloclades) per 4- or 4½-inch pot. Easter cactus is also attractive in larger pots or hanging baskets: pot three or four cells in each 5- or 6-inch azalea pot or 6-inch hanging basket.

Greenhouse space can be efficiently utilized during production. Unlike most crops, Easter cactus can be grown at close spacing from transplanting until flowering without adversely affecting plant quality.

Environmental conditions during growth phase. Easter cactus are not high-light plants. Light shade is recommended from late spring through early fall. Plants should receive full sunlight during winter months in the northern United States. Yellowish or bleached phylloclades indicate excessive sunlight; weak or spindly phylloclades and poor branching occur under low light conditions.

Maintain night temperature at 60 F minimum to maximize growth. Day temperatures should be 5 to 10 F higher.

Preparing plants for flowering. Small, immature terminal phylloclades will not set bud and should be removed from plants before start of flower-inducing treatments. Avoid injuring lower phylloclades when twisting off terminal phylloclades. Terminate fertilization four weeks prior to start of floral induction treatments to discourage new vegetative growth and harden off plants. Resume fertilization at the start of induction treatments.

Requirements for flowering. Flower bud formation in Easter cactus requires an inductive treatment followed by long-day photoperiods and warm temperatures. The inductive treatment may be low temperatures (thermoinduction) or, at warmer temperatures, short-day photoperiods (short-day induction).

- **Thermoinduction.** The standard inductive treatment for Easter cactus is thermoinduction. The night temperature is maintained at 47 to 50 F for four to 12 weeks, depending on the cultivar. Optimum conditions for thermoinduction of Crimson Giant are about six weeks at 50 F nights and 55 to 60 F days. After thermoinduction, plants are grown under long days and 64 to 70 F nights/70 to 72 F days. Use incandescent lamps and either night-break (10 p.m. to 2 a.m.) or day-continuation lighting (sundown to 10 p.m.) to provide long days. Plants will start flowering within 55 to 65 days after start of long days and warm temperatures.
- **Short-day induction.** Our research has shown that short-day photoperiods may be used as an inductive treatment for Crimson Giant. Growers should provide eight-hour short days (using black cloth or similar material) and 64 F nights/70 to 72 F days for six to eight weeks. Afterwards, maintain the same temperature regime and initiate long-day photoperiods as described previously.

A short-day induction treatment will produce fewer flowers compared to a thermoinduction treatment of similar duration, but short-day induction may be utilized during warmer periods when thermoinduction isn't feasible. Crimson Giant has been successfully forced during late spring, summer and early fall using this method.

Light levels are critical. Low light intensities during and after inductive treatments reduce bud count. If cloudy winter weather is present during forcing, high-intensity discharge lighting for 16 hours per day at 400 footcandles will ensure good bud counts.

Scheduling. Easter cactus requires eight to 12 months from propagation to flowering, depending on season of propagation and final plant size desired. Plants propagated in the northern United States during February, March or

Rhipsalidopsis

Red Pride Easter cactus

April will have four to five tiers of phylloclades by early fall and can be flowered for Christmas or later. Plants are sold normally from late January until May and are marketed when flower buds are 3/4 to 1 inch long.

Diseases and pests. Root and basal rots can cause plant losses. Prevention is the best defense. You should use a pathogen-free, well-drained medium. Avoid overwatering and excess fertilization. Periodic drenches with a broad-spectrum fungicide are recommended.

Spider mites and caterpillars will attack Easter cactus. Contact the Cooperative Extension Service in your state for appropriate control measures.

Easter cactus have tremendous potential as a flowering pot crop for North American production. Growers searching for profitable alternative crops should consider Easter cactus for their production programs.

"

Thomas Boyle is assistant professor of floriculture, Department of Plant and Soil Sciences, University of Massachusetts, Amherst. Dennis Stimart is associate professor, Department of Horticulture, University of Wisconsin, Madison.

Research funding was provided by the American Floral Endowment.

Rhododendron, Azalea

Culture Notes

August 1985

by Debbie Hamrick

Azalea
Family: Ericaceae
Genus, species: *Rhododendron obtusum*, *Rhododendron simsii*

Since the mid-1800s plant breeders have been dazzled by the electric flower colors of azaleas brought to Western civilization from the Far East. It didn't take long for the brilliant red, dainty pink, and pure white flowers to work their way into the hearts and greenhouses of floriculturists across Europe and the United States.

Termed the "Cadillac of pot plants" by some, greenhouse azaleas have enjoyed a turbulent love/hate relationship with greenhouse growers over the past century. And, there was good reason. The traditional florists' azalea was a large plant which required two years or more to grow.

But the times have changed. Today's consumer buys flowering plants in a wide range of retail outlets in addition to traditional florists' shops. And, in keeping with the times, the trend in azalea production is toward smaller pot sizes (4½ inches) with a smaller price tag. There is more good news. Crop time for these smaller pots is only one year from cutting to finished product, even less if starting with rooted cuttings. Begin with budded plants and crop time is only eight to 12 weeks—that's quick enough to give any grower's bench a run for its money.

Propagation
Three- to 4-inch terminal cuttings can be taken from healthy stock plants maintained in a protected environment at any time of the year. Shoots for cuttings should "break with a snap." Limber, succulent shoots do not root as readily. Dusting the basal ends with rooting hormone will aid rooting. Direct-stick cuttings into individual peat-type blocks or into flats filled with a peat/perlite or sand mix. Do not crowd. Both mist and bottom heat are ideal for rooting.

In eight weeks cuttings should be well-rooted and ready to transplant.

Growing on, the vegetative stage
Pots: For a small pot program, clay pots are recommended.
Media: Acid-loving azaleas need a media high in organic matter with a pH of 4.0 to 5.5. Good choices include: straight peat, or a combination of three parts

Rhododendron

peat to one part perlite, bark, or peanut hulls. To lower pH, add sulphur prior to planing.

Temperature: Maintain 68 F night temperature and 77 to 86 F day temperature for optimum vegetative growth.

Humidity: 60% relative humidity is ideal.

Light: Maintain a minimum of 2,000 footcandles and a maximum of 4,000 footcandles of light.

Since vegetative growth is stimulated under long-day conditions with many cultivars, a supplemental lighting system, with an intensity of 10 to 20 footcandles, should be used in a year-round growing program. To shorten crop time and keep plants vegetative, provide a continuous 20 to 24 weeks of 16-hour days; natural short days are from September 1 to March 31, so the supplemental lighting should be done during this time.

Nutrition: Feed weekly with a 21-7-7 solution mixed at a rate of 1 1/4 pounds per 100 gallons. Every fourth feeding, supplement the nutrition program math an application of soluble trace elements. For iron deficiency, apply iron chelate.

Water: The fine root systems of azaleas are particularly susceptible to underwatering, overwatering, and overfertilizing. Water stress will retard vegetative growth and promote premature flower bud initiation. With a well-drained media and established plants, azaleas should not be allowed to dry out. Apply 10 percent more water to the pot than the media will hold to leach out excess salts.

Spacing: Depending on the cultivar, space 4 1/2-inch pots two to three per square foot.

Pinching

Pinching shoot tips during the vegetative phase of production promotes lateral branching and increases the number of flowers per plant. In a small pot program, plants should be pinched three times. Pinch for the first time at potting and two more times at 6-week intervals. (During darker times of the year increase the time between pinches to eight weeks.)

Pinch all shoot tips to promote uniformity in subsequent bud production. Use a "soft pinch," or one that removes approximately 1/2 inch of the shoot. Pinching

Rhododendron

off more will reduce the number of lateral shoots produced from the pinch.

After the final pinch, allow plants to grow for six to eight more weeks.

Chemical pinching: Both Atrinal and Off-Shoot-O may be used to chemically pinch azaleas. Each has drawbacks. Atrinal will cause a delay in flowering, and Off-Shoot-O must be applied under very specific environmental conditions. In addition, Off-Shoot-O is not readily available in the marketplace. Follow label instructions for recommended rates.

Budding

Azaleas initiate flower buds when they are given short days, warm temperatures, and/or a B-Nine treatment. Under natural conditions, azaleas develop flower buds during the days of late summer when daylength is shortening and nights are still warm.

To promote flower buds, give plants eight weeks of warm temperatures with a minimum night temperature of 65 F; these warm temperatures should be combined with eight weeks of short days. This combination of warm nights and short days should be done sometime between April 1 and August 31. Eight to 12 weeks after the end of the short-day treatment, flowering will occur.

B-Nine also promotes azalea budding. Apply five weeks after the final pinch. Follow label instructions for rates. Azaleas treated with B-Nine will be smaller than non-treated plants and flowering may be slightly delayed, but, B-Nine will increase the number and uniformity of buds.

Breaking dormancy

Once buds are fully developed, a period of four to six weeks of cool temperatures, 36 to 50 F, is required to break dormancy. This requirement may be met in a refrigerated cooler, cold frame, or unheated greenhouse, depending on the time of year.

Cold frames and unheated greenhouses: Using natural outdoor temperatures to break bud dormancy in azaleas is less expensive than refrigerated cooling, but it is also less precise. Constant temperatures during this phase of production is the key to uniform flowering later on. During the day, temperatures can be moderated using 50 percent shadecloth, and at night, apply enough heat to keep plants from freezing.

Refrigerated cooling: For year-round production, access to a cooler is necessary. Keep Kurume-type cultivars in the cooler for four weeks and other types in for six weeks. If the cooler temperature is kept from 40 to 50 F, give the plants 10 to 20 footcandles of light for 12 hours per day to prevent leaf drop. At a temperature of 35 F lights are not required. Water plants regularly, but do not fertilize. Also, since azaleas are ethylene-sensitive, do not store fruits, vegetables, or other ethylene-producing plants with them.

Forcing

Force plants into flower by maintaining a minimum temperature of 60 F in the greenhouse. Plants will begin to flower in four to six weeks. Higher temperatures will speed flowering, and lower temperatures will slow it.

During forcing, plants may develop by-pass shoots that surround the bud. Using a sideways, twisting motion, remove these shoots when they are 1/2 inch long. By-pass shoots allowed to continue may cause bud blasting.

Diseases

Cylindrocladium scoparium, cylindrocladium disease, causes a blighted or wilted appearance and may affect azaleas at any stage of development. Splashing water spreads spores, and high humidity enhances disease development. Benlate spray or drench may be used for control. Apply at label rates.

Phytopthora spp. and *Pythium spp.* both cause plants to lose vigor. Plants may also have discolored or rotten roots or basal stem rot. These diseases are aggravated by overwatering and cool temperatures. An Aliette, Truban, or Subdue drench may be used for control. Apply at label rates.

Botrytis spp. and/or powdery mildew may be a problem in cool storage in poorly ventilated coolers.

Insects

Gracillaria azaleella, azalea leafminer, causes the tips of new leaves to roll back and turn brown. Damage occurs suddenly. Since the damaging immature stage of the insect is protected by curled leaves, control is difficult. Use Orthene, diazinon, Vapona, or Oxamyl.

Cultivars

There are hundreds of cultivars which may be forced in the greenhouse. In selecting a mix, pick rapid growers and ones with similar dormancy breaking requirements. Each cultivar in the following list has performed well in small pot programs.

Alaska (Rutherford) is white with a chartreuse blotch. Plants may show three types of flowers; single, hose-in-hose, and semi-double. **Dogwood** (Kurume) is pure white with red penciling. Flowers are single. **Gish series** (Rutherfords) feature hose-in-hose flowers. **Gloria** is salmon and white with red spots in the throat. Dorothy Gish is red, and White Gish is white. All are fast growers and have long-lasting flowers. **Mission Bells** (Indica-type) is deep red. The large flowers are semi-double and long-lasting. **Prize** (Indica-type) is rosy-red with blotchy-green foliage. A vigorous grower, the single flowers are long-lasting. **Red Wing** (Indica) is orange-red. Flowers are single and hose-in-hose.

99

Debbie Hamrick is publisher/editor of FloraCulture International *and international editor of* GrowerTalks *magazine.*

Rosa, Mini roses

Culture Notes

February 1986

by Teresa Aimone

Pot roses
Family: Rosaceae
Genus, species: *Rosa* x *species*

Pot roses are beautiful flowering pot plants that are normally grown for Valentine's Day, Easter, Mother's Day, and general spring sales. They can, however, be grown on a year-round basis. Though there are a large number of roses available, there are few that are well-suited for pot rose culture. Miniature roses, floribundas, and polyantha types are normally used for pot culture.

Pot roses can be produced either from dormant bareroot plants or from cuttings. Each technique will be discussed separately.

Production from dormant plants

Most forcers receive grafted, 2-year-old, dormant plants. For Valentine's Day sales, the roses should be potted in late fall (mid-November); Easter pot roses should be potted in late January, and roses for Mother's Day sales should be potted in mid-February. If plants have to be stored for a short period of time before potting, place the plants in a 32 to 35 F cooler. Keep the relative humidity as low as possible.

Once plants have arrived, unpack them immediately and inspect them for any damage. If plants have frozen, you may be able to save them by thawing them slowly in a 35 to 40 F cooler for three to four days. Inspect the plants after this time, and if the canes are still green, the plants are fine. If not, then contact the company who delivered the plants so you can file a claim.

If plants have arrived in fine condition, they may either be immersed in water for a short period of time or wet thoroughly and covered with moist burlap for a day. Either method will encourage new root growth and bud breaks after potting. Most pot roses are grown in at least a 6-inch pot.

The grafted rose plant comes in three grades: there is an "X" grade, given for plants with at least one strong cane; "XX", for at least three strong canes, and "XXX" for at least four strong canes. The XXX grades are preferable since the finished plants will tend to be more uniform and attractive than those plants grown from other grades.

Pot the plants in a well-drained medium. Cut the shoots back to about 6 to 8 inches above the bud union (the bud union is the swollen area on the lower stems from which the canes grow). The pruning will encourage vigorous bud breaks.

Roots may also need to be pruned so the plants will fit into the pots. Prune the roots as little as possible, though. Bear in mind that the fine, fibrous roots help get the plants off to a good start, so if any pruning is necessary to fit the plants in the pot, cut off the heavier roots or the ones that are damaged.

To encourage initial root development, grow the plants at a 55 F night temperature for three to four weeks. Keep plants well-watered and maintain high humidity. If buds are slow to break, the plants may be tented with clear poly for two weeks. This will create a warm, moist environment for the plants which encourages good breaking action. Be sure to raise the sides of the poly on extremely warm days—this will vent excess heat away from the plants.

After the initial 55 temperature, raise the temperature to 60 to 65 F to begin the forcing process. The rate of shoot growth can be controlled by increasing or decreasing the forcing temperature, so manipulating the temperature can be used to time the crop for a specific date. An optimal night temperature is 62 F. Night temperatures may be lowered to 55 F one to two weeks prior to sales—this will aid in hardening the plants and prepare them for long shelf life at the retail or consumer end.

Pot roses are especially popular for holiday and spring sales; they fit well into the production schedule between poinsettias and Easter lilies.

Plants should be pinched when the new shoots are 3 to 4 inches long. Depending on the desired size of the finished product, pinch back to leave either three, four, or five sets of leaves. For Easter roses, pinch the plants no later than six to seven weeks prior to sale. For Mother's Day roses, pinch the plants no later than five to six weeks prior to sale.

Roses respond well to high light conditions. Low light levels—particularly during flower initiation—can seriously affect the number of flowers. If you are in an area of low light, 300 to 500 footcandles of supplemental lighting will improve flower initiation, possibly reduce the number of blind shoots, improve general plant quality, and help prevent flower abortion.

For 6-inch pots, space at either 10 or 12 inches on center.

Roses are not heavy feeders, and recommendations vary widely on how often to feed pot roses. Suggestions range from applying a constant liquid feed of 20-20-20 at 200 ppm, to waiting to feed six weeks after potting and then feeding a minimum of three times. Judge plant growth accordingly when feeding—just bear in mind that pot roses do not require high amounts of feed.

Production from cuttings

Cuttings can be taken from plants that have flower buds just beginning to show color. For best results, select cuttings that have at least five sets of leaves and are a minimum of 1/8 inch in diameter. A rooting hormone will aid in rooting. Soil temperature during rooting should be 70 to 72 F. Recommendations suggest sticking the cuttings directly into 2- or 2¼-inch pots. Follow the same forcing procedures as with dormant plant forcing. Cuttings should be planted three to five per 6-inch pot. Allow 15 to 17 weeks of production time when growing pot roses from cuttings.

Rosa

Miniature roses

Miniature roses are naturally dwarf roses. And they will respond to the same cultural practices used on other pot roses. Miniature roses can grow as tall as 12 to 18 inches, or as short as 5 to 8 inches. Feed miniature roses even less than you would other pot roses; feeding is proportionate to their size.

Varieties

The **Koster series** of polyantha roses are popular. Varieties include **Dick Koster** (deep pink) and **Margo Koster** (salmon). **Mothersday** (deep red) is another polyantha rose. **Sunblaze series** of roses includes **Orange; Scarlet; Royal** (yellow); **Magic** (white with pink edges), and **Pink**.

The **Garnette series** are floribunda-type roses. **Bright Pink Garnette** is a bright deep pink; **Garnette** is garnet-red with a light lemon-yellow base. Other floribundas are **Carol Amling** (deep rose-pink with lighter edges); **Marimba** (pink), and **Thunderbird** (rose red).

Miniature rose varieties worth trying are: **Shy Beauty** (cotton-candy pink); **Wedded Bliss** (semi-double, bright pink); **Orange Sherbet** (bright orange fading to pink); **Per Chance** (coral red with yellow base); **Centerpiece** (deep red); **Added Touch** (coral pink with yellow centers), and **Black Jade** (velvety, dark red.)

"

Teresa Aimone, former editor of GrowerTalks *magazine, is now with Sluis and Groot, Fort Wayne, Indiana.*

Rosemary

Culture notes

August 1988

by Thomas DeBaggio and Thomas Boyle

Rosemary
Family: Labiatae
Genus, species: *Rosmarinus officinalis*

As the Christmas pot plant market becomes saturated with poinsettias, growers might find a unique alternative in rosemary, a linking of Christmas tradition and one of the latest garden trends—herbs.

Rosemary is one of the most important herbs and a multi-purpose species: the leaves, either fresh or dried, are used for culinary flavoring; the aromatic oil distilled from fresh flowering shoots is used for perfume and in medicine; and

in warmer climates the plant is grown as an ornamental evergreen shrub in landscapes. In addition, rosemary makes an ideal pot plant to be enjoyed year-round.

In Europe, rosemary boughs have been used for hundreds of years in Christmas decorations, along with the traditional mistletoe and holly. The sword-shaped, leathery leaves are aromatic, reminiscent of pine or fir needles. Herb specialists have discovered rosemary's usefulness as a decorative centerpiece or as a tabletop tree. As a pot plant, it can be grown in a variety of shapes, from conical "Christmas trees" to topiary standards and living wreaths. How many ornamental pot plants can be used for flavoring the Christmas goose?

Cuttings

Seed: While rosemary can be grown from seed, the process is not recommended for pot plant production. Seed germination is sometimes slow and erratic, requiring 12 to 28 days or more. Optimum soil temperature for germination is 60 F. Light is not necessary for germination. Crop time from seeding to sale in 4-inch pots is 16 to 18 weeks.

Cuttings: Asexually propagated cultivars are best for commercial pot plant production. Take tip cuttings that are approximately 3 to 4 inches long. One-fourth to one-third of the lower leaves are stripped from the cutting prior to sticking. A basal dip in a rooting compound, 0.1 percent indole butyric acid, plus root zone heat enhances the rooting of cuttings and improves uniformity. A well-drained, soilless media (or a mixture of one part soilless media and one

Production stage	Date	Minimum temperature	Spacing
Take cuttings	June 8	70 to 75 F	200 to 1020 flat
Transplant cuttings to 2½-inch pots	June 29	65 F night	Pot-to-pot
Soft-pinch	June 29		
Shift to 6-inch pots	August 10	65 F night	Pot-to-pot
Soft-pinch and shape	August 10		
Soft-pinch and shape	October 12		
Final spacing	October 12	65 F night	10-inch centers
Ready for sale	December 1		

Table 1. Recommended schedule for holiday production of 6-inch rosemary.

part coarse perlite) works well for rooting cuttings. About 200 cuttings can be stuck in a standard 10-inch by 20-inch propagation flat, or cell trays may be used. Rooting medium temperature should be maintained at 70 to 75 F. Cuttings placed under intermittent mist will start to root in 10 days and are ready for transplanting in 17 to 21 days after sticking.

Growing-on

Transplanting: At transplanting, select heavily rooted cuttings and discard weakly rooted cuttings (usually less than 5 percent). For 4-inch pot production, rooted cuttings may be transplanted directly into finishing pots. For 6-inch or gallon containers, cuttings are potted into 2½-inch pots. Plants grow equally well in a soilless or soil-based (one part each of perlite, peat moss and soil) growing medium, but will require more frequent watering in lighter media. Plants are soft-pinched by hand or scissors at transplanting to develop height uniformity, promote upright growth and to encourage branching. Plants grown in 4-inch pots will require about six weeks from transplanting to make a salable, branched plant.

Fertilization: A moderate rate of fertilizer is adequate. Research on rosemary conducted at the University of Massachusetts showed satisfactory growth in soilless media may be obtained by the following fertilizer regimes: constant liquid fertilization at 150 parts per million (ppm) nitrogen using 20-10-20; weekly liquid fertilization at 150 ppm nitrogen using 20-10-20, plus a top-dressing of 4.5 grams (three-fourths teaspoon) Sierra 12-12-15 per 6-inch pot; or a top-dressing of 9 grams (1½ teaspoons) Sierra 12-12-15 per 6-inch pot as the sole source of nutrients. Additions of slow-release fertilizers may be reduced 25 percent when plants are grown in media having more than 20 percent soil.

Temperature and light: As cooler weather approaches, night temperatures should be maintained at 65 F. Day temperatures may be reduced below 65 F to produce stockier plants. Plants will grow in full sun and can withstand greenhouse temperatures of at least 120 F. However, a lightly shaded greenhouse may be used during summer. In spring and fall, plants should be grown in full sun.

Finishing: After six weeks, the plants in 2½-inch pots are shifted to 6-inch or gallon containers, using one plant per pot. At this time, the plants are soft-pinched again. The grower now may begin shaping the plant to attain a conical "Christmas tree" form. Be sure that plants are transplanted upright in the pot and to the depth of the original rootball to avoid a misshaped form. After transplanting, space plants pot-to-pot until final spacing. **Caution:** Too much water may weaken or kill rosemary. It is usually several days before rosemary needs water after transplanting to larger pots.

After eight to 10 weeks, plants are ready for a last (third) shearing and pots are moved to final spacing on 10-inch centers. Five to seven weeks later the plants are ready for sale.

Insects and diseases

Insects: The major pests on rosemary are whiteflies and spider mites. Pesticide labels do not commonly include recommended rates for herbs, and since rosemary will likely be eaten, growers should take care to treat plants accordingly. Soap sprays provide effective control of both pests. Growers should consult extension personnel concerning registered pesticides.

Diseases: Rosemary is subject to root and crown rot. It is aggravated by planting too deeply, growing in poorly drained media and by overwatering. Adequate air circulation will reduce incidence of foliar diseases.

Cultivars

Many asexually propagated cultivars are available, ranging from prostrate, hanging-basket types to upright forms. Scents vary with the cultivar and may be mild or strong. Rosemary flowers are small, clustering along the stem and typically are blue, pink and white cultivars are available.

”

Thomas DeBaggio is owner of Earthworks Herb Garden Nursery in Arlington, Virginia, and is president of T. DeBaggio Ltd., a firm that designs and markets herb plant merchandising aids. Thomas Boyle is assistant professor of floriculture, University of Massachusetts, Amherst.

Saintpaulia, African violet

Culture notes

February 1989

by Russell Miller

African violet
Family: Gesneriaceae
Genus, species: *Saintpaulia ionantha*

For growers, the production of African violets is relatively simple compared to other flowering crops, since pinching, disbudding, photoperiod control and growth regulators for height control are not needed to produce a marketable plant. Without these labor-intensive practices, and because there are new plant forms available to growers, African violets—as a holiday plant or a pot plant for year-round sales—offer a high profit per square foot of greenhouse production.

Propagation is by leaf petiole cuttings, the principle commercial method. African violets can also be started from seed, but only a few cultivars will come true from seed. Because suppliers have moved into producing liners, unspaced and spaced pre-finished, and finished African violets, this pot plant is becoming more economical to produce for many growers.

The petioles on leaf-petiole cuttings should be about ¼ inch long and inserted into the rooting medium without touching each other. A rooting hormone may be used to promote more rapid root formation. Cuttings are rooted in flats at 65 F minimum temperature with no mist, heavy shade and infrequent watering.

After 12 to 16 weeks, 1-inch tall plantlets should be ready to transfer from the mother leaf to a 1-inch liner. After another seven weeks, the plant should be ready to be transplanted into a 4-inch pot. Purchasing pre-finished African violets cuts crop time down to five to six weeks.

The medium should be highly organic and well-drained—80 percent peat moss and 20 percent perlite or styrofoam is good. The medium should be sterile and low in salts with a pH of 5.8 to 6.2.

Watering is a very important factor in producing quality African violets, as overwatering can create crown or root rot. Keep soil uniformly moist, but not wet. Using capillary mats with spaghetti tubes may be preferable because they reduce the number of times the plants must be watered and help prevent underwatering. Overhead watering can be used until flowers are present. Water early in the day and never allow water to stand on the crown or foliage for more than a few hours. If the water temperature is 10 to 15 F warmer or 5 F cooler than the air temperature, cell disruption may occur, which can cause foliage ring spots. Subirrigation or a water temperature regulator should be used if growing African violets on a large scale. If using subirrigation, it's

Saintpaulia

Ballet Olympia

important to avoid salt buildup on the surface of the medium by occasionally applying a clear water soil drench.

Humidity should be kept between 70 and 80 percent, as African violets thrive in high humidity.

Growing temperatures for optimum production vary between 75 and 80 F—although African violets can withstand a minimum of 60 F. Higher air temperatures and direct sunlight should be avoided.

Fertilize young plants on a regular basis with 100 parts per million 14-10-14 liquid feed throughout the crop period. Do not overfertilize as African violets are very salt sensitive.

Light intensity controls flowering. A light intensity of 900 to 1,200 footcandles is ideal. Lower levels may inhibit bloom quantity, while higher levels may restrict growth. At less than 500 footcandles, plants remain vegetative. Cool white fluorescent lamp lighting at 600 footcandles for 15 to 18 hours a day also produces quality plants.

The major insect pests include cyclamen mites and thrips. Cyclamen mites can be controlled with Talstar, Avid, Pentac, Vydate, Oxamyl or Kelthane. Orthene or Mavrik, in use with Resmethrin aerosol, is effective against thrips.

The major disease with African violets is Botrytis, usually caused by high humidity and cool temperatures, and which can be avoided by using a preventative fungicide. Other diseases that may be encountered are crown rot or root rot, usually caused by overwatering and poor soil drainage; petiole rot, likely caused by fertilizer salts that are absorbed by clay pots at the point where the petiole touches the rim; and powdery mildew.

Heating and venting in the evening and avoiding overhead watering can control Botrytis, along with using Benlate or Ornalin. Powdery mildew can be controlled with either a karathane spray or a sulfur burner. Other soil diseases can be prevented with a Banrot drench or with Truban, Banol or Subdue in combination with Benlate.

Available varieties of African violets are increasing in number, providing growers with new colors and plant habits to choose from. There are three important families of African Violets (although all groups include some excellent varieties): the Optimera and Rhapsody series, available from

Saintpaulia

Holtkamp Greenhouses; the Melodie series, available from Sunnyside Nurseries; and the Ballet series, available from Ball Seed Co.

Through recent breeding efforts, the shelf life, the number of flowers and the color spectrum of available African violet varieties have been increased, although breeders are still looking for the ultimate yellow and fire-engine red flower colors. New introductions also include bi-color and miniature pot varieties and, possibly soon, an African violet which blooms continuously.

"

Russell Miller is a former staff writer for GrowerTalks *magazine.*

Thanks to Ron Adams and John Ward of the Ball Seed Co. for their guidiance in the preparation of this article.

Schlumbergera, Christmas cactus

Culture notes

July 1985

by Teresa Aimone

Christmas cactus
Family: Cactaceae
Genus, species: *Schlumbergera bridgesii*
Thanksgiving cactus
Genus, species: *Schlumbergera truncata*

The various types of holiday cacti are often lumped under the common name of Christmas cacti, when, actually, the holiday cacti grown for Christmas sales are usually Thanksgiving cacti. The flowers of both are similar, but leaf margins differ as do natural flowering times; the holiday names associated with these plants indicate when they will naturally flower. The leaf margins on Thanksgiving cacti have a serrated or toothed margin, while Christmas cacti leaf margins are more rounded. Both can be produced from cuttings and require short days for flowering.

Propagation
Mature, single-tier (single segment) cuttings can be direct stuck into either finishing pots or into smaller containers. Drench with a broad-spectrum fungicide after sticking. If space is at a premium at the time of propagation, smaller pots may be more desirable—plants can be shifted later on.

Schlumbergera

Recommendations for the number of cuttings per container are: one per 2¼- or 3-inch; one or two per 4- or 4½-inch, and three per 5- or 5½-inch. If larger containers are desired, adjust the number of cuttings to make certain the finished product will appear full. However, be aware that holiday cacti are fairly low-growing plants, and, in larger containers, the ratio of plant height to container height may look out of proportion. Holiday cacti also make great basket items; plant five or six cuttings per 8- or 10-inch basket. At all initial propagation stages, containers can be spaced pot-tight.

Propagation medium and growing-on medium should have a high organic content, good drainage, and a pH of 5.5 to 6.2. A 1:1:1 peat, perlite, sand or 1:1:1 peat, perlite, and soil medium works well. These are members of the cactus family, and therefore, in general, don't like soggy soil. But they don't like to have dry medium, either. Excessive drying will cause checks in growth at all stages of development, and it's especially detrimental during the bud initiation and development.

For a Christmas crop, propagation can be done from December through late June. Earlier propagation will obviously produce fuller plants since plants can be pinched. For early propagation, take cuttings from December through May 1 to 15. Plants should receive all final prunings or pinchings by the end of June. Any cuttings taken from these prunings or pinches can also be planted. Do not plant any cuttings after the end of June, and do not pinch plants after the end of June if salable flowering plants are desired for Christmas.

Fertilization

Maintain a soil pH of 5.5 to 6.2. Fertilize with 200 to 250 ppm 20-20-20 until bud initiation occurs (early to mid-September). If plants are on a constant liquid feed program and you wish to incorporate slow release fertilizer in your mix, use only ½ the rate of slow release. Holiday cacti respond well to media containing a trace element mix. During the period of fertilization, leach monthly.

Schlumbergera

Light levels

Maintain light levels of 1,500 to 2,000 footcandles during summer and fall; 2,000 to 3,000 during spring and winter.

Flowering

Holiday cacti flowering is controlled by daylength and temperature—short days and low night temperatures are the keys to good bud set and development. For a Christmas crop, plants should be given short days of nine hours beginning September 15—this will cause bud initiation. During this period, night temperatures of 55 to 60 F should be maintained until buds are well-developed. Plants can be given natural daylengths for approximately four to six weeks after the start of short days. Night temperatures above 70 F during short days will either delay flowering or cause plants to revert back to the vegetative stage and no bud set will occur. Flowering will occur at temperatures under 55 F night during short days, but flowering will be sporadic. Do not withhold water while buds are actively forming; this results in undersized flowers or bud abortion. Maintain 65 F nights after the end of short days. Following this schedule, flowering should begin in late November and continue for four to six weeks; blooms can last six to nine days.

Insects

Holiday cacti aren't bothered by many insects; most common are mealybugs and scales. Suggested controls are: Diazinon 50 WP at one tablespoon per gallon; Malathion 25 WP at four to six tablespoons per gallon; Metasystox 25 EC at 1$\frac{1}{2}$ to two teaspoons per gallon.

Physiological problems

The most common problem with holiday cacti is bud drop. This can be caused by one or more of the following: too much or too little water; exposure to cold drafts; presence of manufactured gas (such as ethylene); excessive handling; overfertilization, and a rapidly changing environment. Plants do respond to STS (silver thiosulphate) treatments. STS applications are especially important if plants are being shipped long distances since the buds and flowers are very fragile. Apply STS approximately two to three weeks before shipping, just as the buds are beginning to show color. Spray to run-off with two ounces of STS stock solution per one gallon of water.

Cultivars

Thanksgiving cacti come in a range of colors from white to magenta to red; Christmas cacti are mostly red varieties. Popular Thanksgiving cacti varieties are: **Christmas Charm** and **Christmas Magic** (magenta); **Kris Kringle** (wine-red); **Lavender Doll** (light purple); **Peach Parfait** (peach-red); **Red Radiance** (red); **Twilight Tangerine** (light red/orange), and **White Christmas** (white). **Christmas Cheer** and **Koeninger** are dark red Christmas cacti varieties, and **Norris** is a red variety.

"

Teresa Aimone, former editor of GrowerTalks *magazine, is with Sluis and Groot, Fort Wayne, Indiana.*

Senecio, Cineraria

Culture Notes

July 1986

by Teresa Aimone

Cineraria
Family: Compositae
Genus, species: *Senecio* x *hybridus*

Cinerarias are very colorful pot plants that are inexpensive to produce. Though cinerarias have perhaps lost a bit of the novelty status they once had, they're still much in demand January through May. The daisy-like flowers are available in a wide range of colors and flower patterns.

Propagation
Cinerarias are propagated by seed; plants will flower 28 weeks after sowing. Seed sown in June will produce flowering plants in January. For Valentine's Day cinerarias, sow seed July 15 to August 1. Sowings in late August and early September will produce plants for Easter sales.

Seed are small (150,000 per ounce), so use a fine germination medium. Keep the germination medium moist throughout the germination process. Provide a soil temperature of 70 F. Seed should germinate in 10 to 14 days.

Growing-on
Transplanting: To avoid possible damp-off in the flat, transplant seedlings as soon as they can be handled. Like vinca, cineraria seedlings are very susceptible to damp-off at the seedling stage. Since cinerarias have thin, fibrous root systems, the finishing medium should be lightweight.

Shift seedlings initially into 2¼- or 3-inch peat or plastic pots. After transplanting, drench with Banrot, or a Benlate/Subdue combination. See the labels for rates. As soon as plants become crowded (as soon as the leaves touch), shift plants into their final containers. Cinerarias are commonly grown in 5-, 5½-, and 6-inch containers; smaller pots can be considered for some of the newer, more compact varieties and/or for mass market production.

Fertilization: High fertility levels (especially nitrogen) will cause excessive vegetative growth. Cinerarias lose a great deal of water from their leaves, so the smaller the leaves, the less problem there will be with plants drying out. Feed plants with 100 ppm nitrogen and potassium at every watering; or, feed with 200 ppm 20-20-20 at every other watering. Half-rate of Osmocote (or a similar slow-release fertilizer) may be incorporated per cubic yard of medium.

Senecio

Cindy cineraria

Watering: When plants become pot-bound (as they can quickly do), watering is a major problem, complicated in part by large leaf surfaces. Keep the medium evenly moist, and provide light shade if necessary. Stem rot diseases caused by Pythium and Rhizoctonia can be encouraged if plants are overwatered.

Spacing: A final spacing of 12 inches by 12 inches will provide compact plants.

Temperature: Cinerarias are cool-loving pot plants, and low temperatures are required for plants to initiate flower buds. Maintain 50 to 55 F nights during the growing-on stages. 45 to 50 F temperatures can be used, but growth will be very slow. These latter temperatures could be used if plants need to be held before sales. Plants can be grown warm in the early stages of growth (63 to 65 F), then receive six weeks of 55 F night temperatures to initiate flower buds. After flower buds have initiated, provide plants with eight weeks of 50 to 55 F night temperatures.

Lighting: There usually isn't any problem stimulating vegetative growth on cinerarias; if necessary, however, you can increase vegetative growth by giving plants 18 continuous hours of 10-footcandle light for two to three weeks at the time of transplanting.

Height control: Cinerarias don't normally require chemical height control. B-Nine will limit plant stretching, but it will also delay flowering.

Insects and diseases

Insects: Major pests attacking cinerarias include aphids, red spiders, leaf rollers, whiteflies, and thrips. Temik will provide good overall control. Apply three to four weeks after transplanting, and then every four weeks thereafter.

Diseases: Verticillium wilt and stem rot diseases are common problems on cinerarias. Stem rot diseases can occur from planting the seedlings too deep. For disease control, use good sanitation practices and apply fungicide drenches.

Varieties

Cindy cinerarias were bred for small-to-medium pot sizes. Plants are very uniform, displaying well-matched plant habits and even flowering. Five color shades (blue, carmine, dark red, pink, and copper) and a Mixture are available.

Dwarf Giant Exhibition Mix is a very compact strain of cinerarias with a wide color range.

Fidelio Mixture is recommended for 4-inch pots. Plants have relatively small leaves and a coarse, compact growing habit. The mixture includes the entire color range, plus a fair percentage of zoned flowers.

Jubilee Mix is a small-leafed variety with a wide range of bright colors with and without eyes. Jubilee Mix is an F_1 variety that's good for 4-inch pots.

Nana is a large-flowering variety that is recommended for 6-inch pot production. Nana is a complete mixture of solid and white-eyed flowers.

Sonnet Mixture is an early-flowering strain that performs well in 5- and 6-inch pots. The mix contains both solid colors and bicolors, although the eyed flowers predominate for a brilliant show. Colors include blue, pink, rose, red, and magenta.

Tourette Mixed is good for fast production in 4- and 5-inch pots. Plants have a compact, branched habit. Flowers are $1^1/_2$ inches in diameter.

"

Teresa Aimone, former editor of GrowerTalks *magazine, is with Sluis and Groot, Fort Wayne, Indiana.*

Sinningia, Gloxinia

Culture Notes

August 1987

by Teresa Aimone

Family: Gesneriaceae
Genus, species: *Sinningia* x *hybrida*

Shorter crop times, higher production volumes and lower energy requirements are all desirable qualities of any crop a grower can produce. *Sinningia* x *hybrida* offers growers these advantages and more.

- Plant habit is similar to gloxinia (*Sinningia speciosa*), with a flower type that is between gloxinia and *Streptocarpus*. The novelty flowers provide something new and attractive to consumers, while the plant itself has a habit growers are familiar with.
- Cultural conditions are similar to gloxinia. While cultural requirements are not exactly the same with these two members of the Gesneriad family, only

Sinningia

slight production adjustments need to be made to produce *S.* x *hybrida*. It can be grown under higher light levels and cooler temperatures than gloxinia.
- Prolific flowering. *S.* x *hybrida* could be termed a "multiflora" because of the abundant number of flowers it produces. Consumers get more color per plant.
- Compact plant habit. This pot plant can be grown in 3½- to 4-inch pots, fitting in well with increasingly popular small container programs and also allowing more plants per square foot of bench space.
- Small, flexible leaves. Shipping many members of the Gesneriad family can sometimes be a problem since the leaves tend to be brittle and break during sleeving and packing. *S.* x *hybrida* has soft, supple leaves that eliminate this problem.
- Shortened crop time. Depending on local climate and growing conditions, crop time is four to six months from sow to sell.

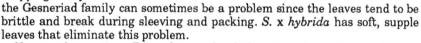

Cultural information

Sowing: *S.* x *hybrida* has approximately 31,250 seed per gram, 885,000 seed per ounce. Sow from late January on for flowering from June on.

Germination: Seeds need a temperature of 71 to 77 F to germinate; they also need light, so do not cover seed. Germination will take place in 14 days. Sow in a light, fine medium that provides good drainage.

Transplanting: Seedlings can be transplanted into flats at 1½-inch by 1½-inch spacing or 2¼-inch pots approximately three to four weeks after sowing. If sown in plugs, plants can go directly from plugs to the final container.

Potting: If seedlings have been given enough space in the germination medium, they can be potted directly into their final container five or six weeks after sowing. This however, can only be done when sown from March on. Plant seedlings slightly deeper than they were in the seed flat. After potting, drench with a labelled fungicide.

Temperature: The optimal growing temperature for *S.* x *hybrida* Diego is 65 to 68 F. Once buds are visible the temperature can be dropped as low as 60 F.

Medium: Use a peat-based medium with extra potassium added.

pH: For best results, soil pH should be 5.5 to 6.

Fertilizer: Do not use a high ammonium-based fertilizer such as 20-20-20. Fertilization requirements are the same as for regular gloxinia production.

Light: Avoid full sun. Optimal light levels are between 2,000 and 2,500 footcandles.

Watering: Like other Gesneriads, *S.* x *hybrida* has a fine root system and is fairly succulent, so they are very sensitive to overwatering. Tube or capillary watering works well.

Crop time: Plants grown during spring production at the optimal temperature mentioned above can be finished in four months from the time of sowing. Production time increases when plants are grown at lower temperatures or at other times of the year.

Pests

Insects: The most common pests on *Sinningia* are cyclamen mites and thrips. Both can be controlled with insecticides or miticides labelled for *Sinningia*.

Diseases: Overwatering, poor ventilation and insufficient spacing are the most common causes of diseases; as a result, Botrytis and/or crown rots can occur. Good cultural practices, as well as preventative sprays of labelled fungicides will help prevent diseases.

Varieties

S. x *hybrida* is available as Diego. Diego includes three colors: Diego Red, bright red flowers; Diego Rose, soft rose, with light throat; and Diego Violet, violet blue flowers.

This novelty, compact pot plant is a crop that fits well into other Gesneriad programs—offering growers a different and interesting item to sell.

"

Teresa Aimone, former editor of GrowerTalks *magazine, is with Sluis and Groot, Fort Wayne, Indiana.*

Smithiantha

Culture Notes

April 1987

by Teresa Aimone

Family: Gesneriaceae
Genus, species: ***Smithiantha zebrina***

With the increased grower interest in 4-inch pot production and the need for new items to satisfy discriminating consumers, flower breeders are continually striving to refine new and old plant material that meets these standards.

Smithiantha zebrina F_1 Jaguar is an improvement over previous *Smithiantha* cultivars. *Smithiantha* is related to gloxinias, and it can be grown together

Smithiantha

successfully with these other members of the Gesneriaceae family. *Smithiantha* is not unknown to greenhouse growers, but previous varieties yielded plants with undesirable qualifies, such as tall growth with non-uniform habit, flowering time, and flower quality.

New to the market, Jaguar produces free-flowering plants, which achieve uniformity in habit, flower color, and flowering time. It is also earlier than older *Smithiantha* varieties; it has orange-scarlet flowers with yellow throats and a distinctive foliage pattern.

Though this new variety can be produced from leaf propagation like other plants in Gesneriaceae, current production is from seed.

Propagation

Smithiantha is a very small-seeded plant. There are 140,000 seed per gram (3,969,000 seed per ounce).

The best time to sow *Smithiantha* is winter through early spring. Depending on the time sown and the production methods followed, *Smithiantha* will flower approximately 4½ to 5½ months after sowing, which allows it to fill in nicely with summer crop production.

Sow seed in a fine, porous peat mixture. Because *Smithiantha* needs light to germinate, do not cover the seed. Supplemental light during the germination process can be beneficial.

Keep the environment surrounding the seedlings humid, but do not saturate the soil. *Smithiantha* seedlings have fine, fibrous root systems, like other Gesneriaceae, so they prefer a moist (not wet) propagation medium.

Maintain soil temperatures of 70 to 75 F. Seed will germinate in 15 to 20 days.

Transplanting

Seedlings can be transplanted approximately seven to eight weeks after sowing. Transplant them into 2- to 3-inch packs or pots, or into a flat using a 2 to 3 inches on center spacing.

Keep the soil uniformly moist to avoid uneven plant growth. Maintain soil temperatures of 68 to 72 F.

Potting

Approximately four to five weeks after transplanting, plants can be moved into their final containers. *Smithiantha* performs best in 3½- to 4½-inch (9 to 11 centimeter) containers. Do not allow the seedlings to crowd each other in the transplanting stage—this will cause plants to stretch and produce undesirable growth in the finishing stages.

Medium

Final growing medium should be a peat-lite mixture that incorporates an additional coarse aggregate, such as sand. Trace elements can also be added to the growing medium.

Maintain a pH of 6.0.

Growing on Fertilization
Feed *Smithiantha* on the same moderate feeding program as gloxinias. Make certain that the fertilizer contains as much nitrogen as potassium.

Temperature
Maintain final growing-on temperatures of 65 to 72 F. The recommended minimum night temperature for Smithiantha is 65 F.

Spacing
Depending on how plant growth proceeds, use a final spacing of 6 inches (14 centimeters) on center. Plants that are spaced too close together will grow tall and have weak stems.

Light
Smithiantha does not like full sun conditions; heavy shading, however, will result in heavy vegetative growth and possible flower shatter. Optimal light levels are 1,350 to 1,800 footcandles (15,000 to 20,000 lux). Long-day conditions will delay flowering somewhat.

The finished product
Grown under these conditions, *Smithiantha zebrina* F_1 Jaguar is sure to produce a profitable crop for the grower and a pot plant that will ensure the consumer lots of enjoyment.

"

Teresa Aimone, former editor of GrowerTalks *magazine, is with Sluis and Groot, Fort Wayne, Indiana.*

Solanum, Ornamental pepper

Culture Notes

May 1985

by Teresa Aimone

Ornamental pepper
Family: Solanaceae
Genus, species: *Capsicum* sp.

Jerusalem cherry
Genus, species: *Solanum pseudo-capsicum*

Ornamental peppers and Jerusalem cherries (also known as Christmas peppers and Christmas cherries) are often grown together since they have the same basic cultural requirements and are marketed at the same times of the year. Though their names imply that they are grown for Christmas sales, ornamental peppers and Jerusalem cherries can be produced for sale from September to February. Areas with milder climates can use these two crops for late season bedding, too.

Christmas peppers are normally characterized by having tapered, conical fruit; fruit on individual plants can change from purple to cream to red or orange in one growing season. Jerusalem cherries have spherical, red or yellow fruit. The seed of ornamental peppers are slightly larger (with 9,000 seed per ounce) than those of Jerusalem cherries that have 12,000 seed per ounce.

Propagation

These two plants are most often grown from seed, though they can be started from cuttings. Seed should be sown on a 1:1:1 soil, peat, perlite or calcined clay media; the media should have good water-holding capacity or the seedlings will have a tendency to stretch. Both ornamental peppers and Jerusalem cherries will germinate in approximately 15 days at 70 F soil temperature. A light level of 2,000 to 4,000 footcandles is recommended; under high light conditions, maintain the recommended footcandle levels to protect seedlings and possibly increase germination.

Ornamental peppers should be sown from late April to early May for a Christmas crop; Jerusalem cherries require an earlier sowing in late February and March for Christmas production. Later sowing will provide attractive potted plants for post-holiday sales. Producing either peppers or cherries from cuttings will cut 2½ to three months off of the crop time.

Solanum

Cultivation
Seedlings should be transplanted into 2¼-, 2½-, or 3-inch pots as soon as seedlings are large enough to handle. The same media used for germination is fine for finished plants. If a 4-, 5-, or 6-inch finishing container is desired, shift in May; do not allow plants to become rootbound. After this shift, set the plants outside. Setting plants outdoors ensures that pollination by bees or wind will occur, and plants will have good fruit set. Bring plants indoors in September and grow-on at 50 F nights. If plants have been produced up to this point in 4- or 5-inch, shift to 6-inch pots after plants have been brought indoors, if desired.

Pinching: During the growing period from April to June, plants should be pinched periodically. Pinch when there are three to four nodes of new growth. After July 15, do not pinch plants anymore. Many varieties are fairly self-branching and do not require pinching.

Feeding and watering: High fertility levels cause excessive, stretched growth. Low nitrogen levels are necessary, and a constant liquid feed of 100 ppm 20-20-20 is sufficient. Spurway soil test maximum readings should yield 20 to 25 ppm N per week; four to six ppm P per week, and 25 to 30 ppm K per week. Both ornamental peppers and Jerusalem cherries require frequent, heavy waterings. As with most flowering crops, feed and water should be reduced as plants get closer to selling times.

Handling: Plants are sold as fruit color develops throughout the fall. Care should be taken during shipping and handling since the fruit is easily broken from the stems.

Insects and diseases
Common greenhouse insects include aphids, red spiders, and mealybugs. Since the plants are grown outdoors for a portion of their life cycle, pests such as grasshoppers and worms can also be a problem. Common greenhouse

Solanum

insecticides will eradicate the "indoor" pests. Dipel or Thuricide will get rid of pests found in outdoor conditions.

Ornamental peppers and Jerusalem cherries require copious amounts of water, so be careful of stem rot and root rot diseases. Lesan or Terraclor can be used for control.

Variety

Ornamental pepper and Jerusalem cherry varieties are continuously being bred to produce more compact plants with better basal branching and more vibrant colors. All of the following varieties except Red Giant are ornamental pepper varieties.

Holiday Time is an All-America Selections winner. Red fruit are borne on upright, early-flowering plants.

Holiday Cheer has spherical fruit—unusual for an ornamental pepper. Plants have good, compact habit; fruit are initially yellow then turn to red.

Fireworks have a more spreading growth habit. Conical fruit start out cream-colored then turn red; they're borne on early, prolific plants.

Fips bears clusters of 1-inch fruit high above the foliage. Fruit mature red, starting out white then turning to yellow.

Fiesta has long, slender fruit. Initially, fruit are white then mature to red.

Candlelight, another All-America Selections winner, has 1½-inch tapered fruit that start out green then turn to red. Plants are compact and prolific' Candlelight is very good in warm climates.

Red Missile is the first F_1 hybrid ornamental pepper. The mature, red, tapered fruit are white at fruit initiation, then turn yellow.

Holiday Flames is similar to Fiesta, but the fruit are thicker. Plants are early and prolific and produce fruit that turn from cream to fiery red.

Ball Christmas Pepper has cone-shaped fruit that are first green, then purple, then red. Plants are compact and bear lots of fruit.

Aurora doesn't need pinching if plants are transplanted and shifted on time. Fruit changes colors from purple to cream to red.

Red Giant is a good variety of Jerusalem cherries. These 10- to 11-inch tall plants have a 9- to 10-inch spread and bear large, round, orange-red fruit.

"

Teresa Aimone, former editor of GrowerTalks *magazine, is with Sluis and Groot, Fort Wayne, Indiana.*

Culture Notes

September 1987

by Lin Saussy Wiles

Cape Primrose
Family: Gesneriaceae
Genus, species: *Streptocarpus* x *hybridus*

Streptocarpus, or cape primroses, are more than beautiful greenhouse plants and have recently gained greater attention for their many uses. They make enduring pot plants with the newer hybrids much easier to grow than their African violet relatives. The cut flower industry should take note of these plants because they produce long-lasting, orchid-like flowers, some lasting as long as three weeks. Interiorscapers should look at them as a color accent in their designs, since they tolerate lower light levels than many flowering plants. Some can even be used as bedding plants in shady northern gardens.

Propagation
Vegetative: Divide plants using a sterile, sharp knife to separate the plant into as many sections (immature leaves surrounded by large mature leaves) as there are available. Most mature plants will have at least two or three sections. Some types require a fungicide drench; or the cut surface can be air-dried for a few days prior to packing soil next to the cut.

To prepare cuttings, select vigorous mature growth, neither oldest nor youngest leaves, and take longitudinal leaf sections or leaf cuttings. Longitudinal leaf sections are prepared by cutting the two sides of the leaf away from midrib, then placing the cut sides down in moist vermiculite.

Leaf cuttings involve cutting the leaf into a series of wedges with the point facing the bottom of the leaf and inserted into the moist vermiculite. Some cultivars root well without the use of rooting hormones. Bottom heat at 70 to 75 F is beneficial and leaves should be kept moist. Detach and move each plantlet to a 2½-inch pot when roots are 1 inch long. Some variations can arise from leaf propagation since *Streptocarpus* are capable of producing plants from a single cell.

Seed: *Streptocarpus* seed are extremely small (900,000 to 1,800,000 seeds per ounce) and should be sown in a very fine particled mix such as peat/vermiculite. Sow at any time, being careful not to cover seed since they require light to germinate. Place the seed flat in a warm place (68 to 72 F for best germination) out of direct sunlight. Mist or bottom water the flats to keep the soil moist but not soaked. Seed is also adaptable to growing in plugs. Seed germinates in 10 to 20 days. The sowing date will affect the time required for plants to flower with an additional month or more needed for overwintered crops.

Growing on

Transplanting: Plants can be transplanted four to eight weeks from sowing when plants are ¼ to ⅜ inch wide. Transplant into 2½-inch pots using a well-drained peat-lite medium with pH of 5.6 to 6.0. A highly porous medium ensures adequate drainage and helps keep roots cool. Cooler roots aid plants' ability to withstand temperatures over 85 F. Seedlings grow slowly initially; in six weeks, or more, the plants can be transplanted into 4-inch pots and spaced on 6- to 8-inch centers for smaller hybrids or 10- to 12-inch centers for larger types. Plants can be held in 2½-inch pots until budded, or first flowers

appear, prior to transplanting. This tends to reduce leaf growth with little or no reduction in flowering. Once plants are in 2½-inch pots, clones and seedlings are handled the same way,

Lighting: Tolerances vary sharply among cultivars, but the optimum range would be 1,000 to 3,000 footcandles of light. Plants can be maintained from 150 to 6,500 footcandles but flowering generally occurs between 500 and 4,200 footcandles. Peduncle length varies inversely with the light level; higher light produces shorter stemmed flowers.

Temperature: Most plants decline in high temperatures, although newer hybrids can tolerate a range of 45 to 105 F. Optimum summer temperature is 60 F nights with soil kept evenly moist. Optimum winter temperatures are 50 to 55 F nights with soil kept slightly drier. Daytime temperatures ranging to 80 F are optimum. Temperatures over 90 F at bud set tend to delay flowering. Good ventilation helps plants withstand hot temperatures. For less heat-tolerant types, roots can be cooled by adding moistened long-fiber sphagnum moss to the top of soil in pots as a mulch. Peduncle length varies inversely with temperature level; high temperature produces shorter flowers and smaller leaves.

Fertilization: After transplanting into 2½-inch pots, begin constant feed using 15-30-15 at 75 to 100 ppm N at every watering. When using a soilless mix, a nitrate nitrogen source will produce good plants with top grade flowers. Ammoniacal nitrogen produces weak plants that have flabby, soft flowers with poor vase life and weak stems. To keep plants more compact, fertilize once a week.

Watering: Let soil surfaces dry between waterings, then water thoroughly. Avoid heavy watering until roots fill the pot. If plants dry too much leaves will wilt, but plants recover quickly once watered.

For plants grown on capillary matting, water overhead initially to establish soilless mix-mat contact. Mineral soil should not be used in the mix—the air spaces will be blocked with capillary water.

Most varieties have a resting phase for two to three months after flowering; others will have a resting phase if stressed, and should be watered only enough to prevent waiting until flowering resumes.

Pests

Insects: *Streptocarpus* are not usually afflicted by insects, but if mealybugs or aphids should occur, treat plants with Orthene or Safer's Insecticidal Soap.

Disease: Crown rot and Botrytis can occur but are uncommon. To prevent crown rot, improve soil drainage, avoid overwatering and use sterilized soil. For control of crown rot use sulfur, Captan, Benlate or Banrot. Botrytis can be prevented by proper plant sanitation and repotting on schedule. For control of Botrytis use Benlate.

Cultivars

Holiday Hybrids (F_1 hybrid series bred by Park Seed) are tolerant of temperature and light extremes. They are early flowering, everblooming, uniform and have long-lasting 2½- to 3½-inch flowers.

Concorde (bred by Floranova) is the first F_1 hybrid series. Plants are early flowering and show good uniformity.

Nymph series (developed In England). These clones have good quantities of small flowers through most of the year.

Weismoor Hybrids (bred by Carl Fleischmann of West Germany) are the first seed-grown plants with uniform flower type. Flowers vary in size with some 4 to 5 inches across.

99

Lin Saussy Wiles was a plant breeder with Park Seed, Greenwood, South Carolina when this article appeared.

Trachelium

Culture notes

July 1988

by Lewis Howe

Throatwort
Family: Campanulaceae
Genus, species: *Trachelium caeruleum*

Throatwort, or *Trachelium,* is native to the Mediterranean. This herbaceous perennial produces large, corymbose heads of tiny violet blue or white flowers. The plant name is derived from *trachelos,* meaning "neck" and refers to the plants reputed value for relieving diseases of the throat; hence the common name throatwort.

Trachelium has received considerable interest from many seed companies as new pot and cut flower crops enter the market. *Trachelium* grows well and is floriferous when grown in pots, providing a fresh look in cut flower arrangements. While crop time is longer than many pot and cut crops, *trachelium* is economical to produce because it prefers a cool growing environment.

Propagation

Seed: *Trachelium* seed is available from several companies. Seed is fine, containing 2 million per ounce (70,550 per gram).

Trachelium can be propagated by cuttings, however this method is seldom used. Thinly sow seed evenly into germination trays containing a peat-lite medium. Do not cover seed. Place flats in a well-ventilated area. Supplemental lighting speeds germination and produces vigorous seedlings. Seed germinates at soil temperatures between 55 to 60 F in 10 to 15 days. If soil temperature exceeds 70 F, germination is severely inhibited.

Although a perennial, *Trachelium* can be treated as an annual or biennial. To grow as an annual for flowering plants in late summer, sow seed from February to April. Pinch plants once for branched habit or leave unpinched to produce a single stem. To grow as a biennial to flower the following summer, sow seed in July. When plants are 3 to 4 inches tall, pinch the terminals to promote branching. Pinch the apical tips when axillary branches grow 3 to 4 inches. Do not let the plants get root-bound during the growing period in late summer and fall. Transplant into 5- or 6-inch pots and establish before overwintering in the greenhouse. Plants from spring sowings will be shorter than plants from summer sowings.

Plugs: Seed germinates in 10 to 15 days at 55 to 60 F soil temperatures. Seed is very small and when grown in small plug cells (½ inch or less) produce transplantable plugs in seven to eight weeks after sowing. After germination,

grow seedlings at 65 to 70 F. Maintain uniform soil moisture during early development. Apply liquid fertilizer, such as 20-10-20 at 65 ppm nitrogen, twice per week beginning at the first true leaf stage. The production of salable 4-, 5- or 6-inch pot plants is accomplished in four and a half to six months. Plugs can reduce crop time by seven to eight weeks.

Cut flowers

Trachelium adds a distinctive accent to flower arrangements with dark green, serrated foliage and globus flower head characteristics. Plants are grown in pots or beds in the greenhouse or field. Each plant produces two to three large blooms and five to seven smaller blooms at first flower. More large blooms are produced when the first, or center, bud is pinched. Plants pinched one to two times while vegetative produce axillary breaks, resulting in more blooms. Year-round production is possible in the greenhouse by using supplemental light to extend daylength and warm day temperatures. From seed to first flower, production time takes five to six months in the winter. Cut flowers harvested at bud stage last 10 to 14 days.

Growing on

Transplanting: For flats, transplant into 804 or 806 inserts seven to eight weeks after sowing. Allow four to five weeks to establish in market packs and sell green with well-illustrated picture labels. For pot production, transplant established plants from marketpacks to 4-, 5- or 6-inch pots. One plant per pot produces a quality pot plant which can be cut. Space plants on 6- to 8-inch centers to avoid stretching.

Light: Grow under long days, 12 to 14 hours of full sunlight. Winter production is possible when plants are grown under long day conditions.

Temperature: During the active growing stage, grow at 60 to 65 F day temperatures and 50 to 55 F night temperatures. Plants produce strong flower

Trachelium

stems at these temperatures. If seed is sown in July to flower the following summer, grow at night temperatures of 40 to 45 F and day temperatures of 50 to 55 F.

Fertilization: *Trachelium* is a moderate feeder. If grown in a greenhouse in a peat-lite medium, begin a constant liquid feeding program using 20-10-20 formulation at 200 to 250 ppm nitrogen. If weekly applications are preferred, apply 350 ppm nitrogen. If plants are held in a cool greenhouse, especially during winter, bimonthly applications, are needed to maintain strong root systems during the overwintering period. Apply a 10-30-20 formulation at 75 to 100 ppm nitrogen.

Growth regulators: If plants are grown cool and in full sunlight, no growth regulator is required.

Insects: *Trachelium* may be attacked by several greenhouse pests, namely aphids and spider mites. Monitor and scout growing areas daily. Early detection of infestations is important so terminals and foliage are not damaged. As with perennials, many plants are not listed on chemical labels. Consult label restrictions before applying.

"

Lewis Howe is a research assistant with Park Seed, Greenwood, South Carolina.

Torenia

Culture notes

May 1988

by Meredith Shank

Torenia
Family: Scrophulariaceae
Genus, species: *Torenia fournieri*

Sometimes the most interesting new plant is one that's been around awhile. New developments in torenia cultivars and their culture are bringing renewed interest in this class from bedding and pot plant growers. *Torenia fournieri* is most commonly known as the wishbone flower, for the wishbone formed by the fusion of two stamens in each flower. Another lesser-known common name is Florida pansy, as it is a non-fading substitute for pansies, which can lose their color in southern climates. Individual torenia flowers resemble those of the snapdragon, a member of the same family, but torenia has a subtle charm all its own.

Torenia is versatile. It does best in cool semi-shade, but will tolerate a much wider range of conditions—from just a few hours of morning sun to full sun in many areas, if moisture is adequate. Even the leaves are attractive: glossy green, which become bronzed as the cool nights of fall arrive. Although it's not the first flower to bloom in the garden, it makes up by looking good all summer, right up to frost. Newer, compact cultivars make excellent 4-inch flowering pot plants that last well on a sunny windowsill indoors. And torenia is a quick and easy greenhouse crop to produce.

Propagation

Seed flats: Torenia seed is fine, at 405,400 seed per ounce. Sow in rows in a sterile, peat-lite medium. Do not cover. Germinate at 70 to 75 F soil temperatures. Seed germinates in flats in 10 to 14 days. Giving the seedlings 18 hours of light at 750 footcandles at bench level until transplanting will hasten flowering. Fertilize at 50 ppm with 15-16-17 Peat-Lite Special.

Transplant to cell-packs or 4-inch pots in three to four weeks.

Plugs: We sow torenia on an Old Mill Seeder, using the petunia setting. On a Blackmore, use the #2 rubber tip manifold. We recommend germinating the seed in a growth chamber, at 70 to 75 F soil temperatures. Remove from the chamber at early emergence, usually six days. Lighting for 18 hours at 750 footcandles will hasten flowering. Fertilize in the plug tray with 50 ppm 15-16-17 Peat-Lite Special as constant feed. This feeding will also make an attractive salable plug with good dark green foliage. Transplant into cell-packs or pots at four to five weeks.

Growing on

Transplant: Transplant into a well-drained sterile mix. For pot production, one seedling per 3- to 4-inch pot is sufficient, three to four seedlings to a 6-inch pot. Use a soilless mix with a pH of 6.0 to 6.2.

Light: Maintain high light levels when growing on. Full light intensity will result in compact plants, even where temperatures are warmer than the ideal. Torenia is non-photoperiodic, and can be flowered year-round.

Temperature: Torenia can be grown cool; 58 nights and 62 days are ideal. It can be grown at warmer temperatures in high light areas and still remain compact, but growing warm in low light will cause stretching.

Fertilization: Torenia is a heavy feeder. Provide a constant feed of 200 ppm N in a 15-16-17 formulation. A Peat-Lite Special formulation with trace elements is ideal. Proper fertilization enhances the green leaf color for a more attractive plant at point-of-purchase.

Height control: When growing torenia as a pot crop, a soft pinch and adequate light provides the best height control. A soft pinch two weeks after transplant will improve branching. In low light areas, a second pinch may be helpful when 1½ inches of new growth is visible, but will add five to seven days to crop time, and is usually not needed. Cycocel and B-Nine have limited effectiveness. New dwarf cultivars are the best bet for 4- or 6-inch pots.

Pest and diseases: Torenia may be affected by common greenhouse pests such as whiteflies and aphids. They are usually controlled by standard spray programs. No phytotoxicities have been noted in trials. For control of powdery mildew, grow in a well-ventilated environment, and watch for excess water and humidity. New cultivars appear to be less susceptible than older strains.

Varieties

Nana compacta: Traditionally grown variety, used primarily for bedding. Blue flowers with dark blue markings.

Compacta Blue: Traditional bedding plant variety Blue flowers with a yellow and violet blotch.

Clown Mix: A 1989 All-America Selections winner. A new strain with colors including dark blue on white, rose on white, and white with a touch of yellow, as well as pastel and dark blue shades. Improved compact habit makes it excellent for 4-inch pots as well as bedding plant packs.

"

Meredith Shank is with Pan American Seed Co., West Chicago, Illinois.

Viola

Culture notes

June 1990

by Jim Nau

Pansies and violas
Family: Violaceae
Genus, species: *Viola* x *Wittrocklana*, **Garden pansy**
Viola cornuta, **Horned violet**
Viola tricolor, **Johnny jump-up**

Pansies and violets have been around for decades but it is only in the past 10 years that the market has increased so dramatically for these plants in the bedding plant arena. Violas as a group come in all the main colors plus shades in between, perform their best under cool (60 F or less) temperatures and are available in both blotched and unblotched faces. While culture for all three types is very similar, some differences do exist.

Garden pansy (20,000 seeds per ounce)

This short lived perennial is most often grown as an annual. Garden pansies germinate in 10 days or less when given 65 to 75 F and seed is lightly covered with medium. Upon germination, drop temperature to 58 to 60 F until transplanting seedlings. Transplant seedlings 15 to 20 days after sowing; adjust growing-on temperature to 50 to 55 F once roots have started to develop in the media. From a December or January sowing, it takes 14 to 15 weeks to flower in 32 packs when grown at 50 to 55 F. For flowering 4-inch pots, allow 16 to 17 weeks with two plants per pot to fill it out.

In the home garden, pansies planted by the middle of May will flower and look their best until around the first week of August. Plant in full sun to partial shade and space 8 to 10 inches apart. Plants grow to a height of 6 to 8 inches. Pansies are tender perennials that may survive mild winters. In regions with severe cold, pansies do best when covered for the winter, although there is no guarantee plants will come back in spring.

In the southern United States allow 11 weeks for flowering packs; 14 weeks for flowering 4-inch pots. In the garden, plant from late summer to December for blooming plants through June. Space 8 inches apart in the garden in the South; 8 to 10 inches in the North.

Varieties available in garden pansies differ by genetic makeup, flower size and blotched versus unblotched faces. Genetically, pansies are available as F_1, F_2 or open-pollinated strains. F_1s are the most heat tolerant and have the longest color season. The F_2s and open-pollinated strains get leggy as heat

Viola

increases, but they also offer the most unusual flower colors of the old Swiss Giant types.

Among the most popular F_1 series, the Universal, Crystal Bowl and Maxim varieties are the best of the 1½- to 2-inch flower size types on the market today. These varieties are available in a wide range of colors plus a mix and put on a full season show until hot weather. The Crystal Bowl series is a plain faced or unblotched variety, while all the Maxim varieties have a prominent, dark blotch except for Sherbet, which has a rose colored blotch. Finally, the Universal series is made up of both blotched and unblotched varieties.

In the medium flower size category of 2 to 3½ inches the Crown and Roc series are two of the leading garden performers. The Crowns have unblotched flowers in a number of colors while the Rocs are predominantly blotched, except for Golden and Orange, which are nonblotched.

The largest class available, the Majestic Giant series, is the most popular variety on the market. Characterized by flowers of 4 inches or more across, Majestic Giants are one of the few pansies on the U. S. trade that are asked for by name at garden centers.

I have noticed a new entry into this class over the past two years that rivals the flower size of the Majestics but on plants that support the flowers better. This series is called Happy Face and has a number of separate colors plus a mixture. At present, it is available in limited amounts this season and will become more visible during the next several years.

Horned violets (24,000 seeds per ounce)

This class of violas is the least popular of the group and is sold most often as mixes. Characterized by having small flowers of 1 to 1½ inches, the plants look very similar to garden pansies, except for the smaller flower size. In cropping, treat horned violets the same as garden pansies. This class flowers in cell packs in 12 to 13 weeks.

All of the commercial varieties available on the trade are predominantly open pollinated. This means they do not have the strong flowering associated with F_1 hybrids. This class is excellent, however, in either an annual or perennial border and can be used as garnishes in salads.

Johnny jump-ups (45,000 seeds per ounce)

Johnny jump-ups have the smallest flower size of all violas sold in the United States today. Blooms are predominantly ¾ to 1 inch across, and plants are widely sold for use in perennial gardens. Johnny jump-ups are the primary viola used as an eatable blossom that adorns plates in finer restaurants. As a group, the jump-ups are more popular than horned violets in the trade, although the majority of sales are in only two varieties.

As for cropping, use the culture provided under pansies. Jump-ups flower in 11 to 12 weeks in packs and are the most vigorous of the violas. Growing too warm causes stretching. Grow these plants no greater than 55 F while they are in cell packs.

Two varieties are commonly available. Johnny jump-up itself is a three-colored variety of violet, lavender and yellow. It is sold under this name or under the name of Helen Mount. The other variety is a solid purple with a yellow eye that is sold under a number of names, with common ones being Blue Elf, Prince Henry and King Henry. These are excellent performers and also well-known by the consumer.

"

Jim Nau is trials and product development manager for seed at Ball Seed Co., West Chicago, Illinois.

Zantedeschia, Calla lily

New Zealand Callas

February 1988

by A. C. Jamieson

Directions for today and tomorrow

New Zealand callas are an exciting new flower crop with tremendous potential as a cut flower, pot plant or garden plant. In the continual quest to produce a better calla and answer the many production questions, New Zealand producers, growers and marketers are funding extensive calla research.

This commitment has come from producers of the New Zealand calla who recently formed the International Calla Association (N.Z. Inc.) The association coordinates and funds research in calla production, initiated an independent cultivar breeding and evaluation program, coordinates promotion and marketing of New Zealand callas, and has established tuber grades and quality standards.

Fine tuning production through research

Research is a major activity with the ICA, initiating and funding investigations into disease prevention and control, storage methods, growth regulators, preplant treatments, plant growth and environmental control. This work is being carried out in New Zealand and the US.

Results of all ICA-funded research remains confidential to its members for a specified length of time before being released publicly.

Disease prevention: Bacterial soft rot of both tubers and cut flower callas can be a major problem and is the subject of considerable research effort.

Zantedeschia

Areas being investigated include:
- **The cause.** Both Erwinia and Pseudomonas bacteria have been isolated from diseased material and it is uncertain which is the primary causal organism.
- **The infection process and how the disease enters the tuber.** Bacteria are unable to breach intact plant tissue and must enter through natural openings or wounded tissue.
- **Host susceptibility.** Some calla hybrids appear to be more resistant to soft rot bacteria and can be selected for use in plant breeding programs.
- **Pathogen survival.** Pathogen survival time in the soil and the relationship between the bacteria and tubers will affect control programs.
- **Soft rot control.** This research is aimed at the *in vitro* screening of bacteriocides for effectiveness against the causal organism(s), and the control of soft rot on intact tubers after lifting. Tuber immersion and/or spraying the bacteriocide have been trialed to determine the most effective method of application.

Storage methods: The ICA is funding research into long-term tuber storage to determine the best temperature, humidity and storage time. This research is being evaluated on the basis of top quality flowers.

Research is also under way to determine the interaction between plant growth regulators and storage. This will indicate if flowering percentages can be enhanced while tubers are in storage.

Further research is comparing the effect of fluctuating and constant tuber storage temperature. This work will indicate the best treatments for performance and flowering control.

Results so far indicate that tubers can be stored for long periods without deterioration. Further work is planned to fine-tune calla tuber storage requirements so that greater flowering control can be achieved.

Growth regulators: Early research has shown that plant growth regulators, particularly gibberellic acid, can significantly increase flowering in cut flower and pot plant calla varieties. GA has been blamed for some flower distortion and recent research has looked at alternative growth regulators.

The method of applying GA also requires investigation. Tubers have traditionally been immersed in GA but the risk of disease spread has meant that spraying tubers may be a viable and safer alternative.

Researchers have investigated the use of GA in combination with gibberellin antagonists such as Bonzi. Bonzi, with its ability to dwarf plants, and GA with its ability to increase flowering, have meant that initial guidelines have been devised for successful pot plant production.

In New Zealand, scientists are testing these guidelines on a wide range of calla pot plant hybrids, which will lead to firm treatment recommendations.

Plant growth and preplant treatments: The rapid development from tissue culture to flowering-sized tubers requires a detailed understanding of how the plant grows. This allows refined plant management procedures to be implemented for the development of a high quality tuber.

In a recent trial, ex-flask tissue-cultured grade 6 tubers (0.4 inch or 3 centimeters minimum circumference) were grown under ideal environmental conditions. Destructive harvesting was carried out at regular intervals, and analysis has detailed tuber growth patterns. This has highlighted timing for cultural practices such as irrigation and nutrition for maximum tuber growth.

Environment control: Calla flowers appear to fade when grown in greenhouses. Possible causes include day and night temperatures, the extent of temperature fluctuations, and light intensity. Environmentally controlled greenhouses are being used to study the effects of differing day/night temperatures and varying light intensity in an attempt to answer these questions.

Zantedeschia

On the road to new varieties

Long term research must focus on continual plant breeding to produce exciting new callas for future markets. The ICA has already initiated a breeding program by selecting New Zealand callas currently available and standardizing their production. Qualified plant breeders will use this gene pool to select and breed superior calla hybrids. The criteria for selection will be dictated by the plant breeder, the grower and the market place.

Independent cultivar evaluation of hybrids suitable for cut flowers, pot plants and garden plants is under way under the supervision of New Zealand plant scientists.

Creating the worldwide blueprint for New Zealand Callas

ICA membership gives calla producers and marketers access to research results and updates every two to three months. It also gives inside information on product development, a say in the future direction of research, marketing and promotion benefits, the chance to expand national and international business contacts, and access to experienced international research workers and growers with a pool of practical and scientific information.

The main thrust of the ICA-sponsored research is to provide a production blueprint for the New Zealand calla and ensure that continued breeding keeps the New Zealand calla in the forefront of world production as a new and exciting flowering plant.

99

A.C. Jamieson is senior horticultural consultant, Ministry of Agriculture and Fisheries, Palmerston North, New Zealand.

'GrowerTalks' Bookshelf

Key horticulture texts for professional/commercial growers

Ball Culture Guide: The Encyclopedia of Seed Germination
Ball RedBook 15th Edition
CO_2 Enrichment in the Greenhouse
The Commercial Greenhouse
Complete Manual of Perennial Ground Covers
Dictionary of Plant Names
Foliage Plant Production
Fresh Cut Flowers For Designs
Gerbera Production
Greenhouse Engineering
Greenhouse Operation and Management
'GrowerTalks' on Crop Culture
'GrowerTalks' on Plugs
Herbaceous Perennial Plants
Hydrangea Production
Introduction to Floriculture
The Miller Report
The National Arboretum Book of Outstanding Garden Plants
Nursery Management: Administration and Culture
Ornamental Horticulture (McDaniel)
Ornamental Horticulture: Principles and Practices
Plant Pathology: Principles and Practices
Plant Propagation: Principles and Practices
Plant Science: Growth, Development and Utilization
　of Cultivated Plants
Practical Horticulture: A Guide to Growing Indoor
　and Outdoor Plants
Profitable Garden Center Management
The Standard Pesticide User's Guide

'GrowerTalks' BookShelf . . . a commitment to education

P.O. Box 532, Geneva, IL 60134-0532
(1) 708-208-9080   FAX (1) 708-208-9350

GrowerExpo ᔆᴹ

Sponsored by 'GrowerTalks' magazine

GrowerExpo is a four-day program that features seminars and a trade show tailored specifically to the needs of the greenhouse grower. Greenhouse growers, owners and managers have the opportunity to learn about the newest technology and crops, along with business and management.

GrowerTalks' staff sees **GrowerExpo** as a forum for exchange of ideas and information and an opportunity for industry leaders to make valuable contacts. **GrowerExpo** meets the needs of commercial greenhouse growers and industry suppliers by providing professional programs targeted to greenhouse businessmen.

GrowerExpo is committed to helping growers solve problems—tuning into the issues facing growers, owners and managers.

Growers talking to other growers continues to be the driving force behind **GrowerExpo**. Count on us to keep you informed.